Publishing

ICE Guide to Careers in Civil Engineering

Institution of Civil Engineers

Published by ICE Publishing, One Great George Street, Westminster, London SW1P 3AA.

Full details of ICE Publishing representatives and distributors can be found at:
www.icebookshop.com/contact.aspx

Other titles by ICE Publishing:

Initial Professional Development for Civil Engineers, Second edition
Patrick Waterhouse with H. Macdonald Steels. ISBN: 978-0-7277-6098-2

Civil Engineering Procedure, Eighth edition
The Institution of Civil Engineers. ISBN: 978-0-7277-6427-0

Successful Professional Reviews for Civil Engineers, Fifth edition
Patrick Waterhouse. ISBN: 978-0-7277-6609-0

www.icebookshop.com
A catalogue record for this book is available from the British Library

ISBN 978-0-7277-6667-0

ICE Publishing is a division of Thomas Telford Ltd, a wholly owned subsidiary of the Institution of Civil Engineers (ICE).

Cover photo: Victoria Square Park, Belfast

Commissioning Editor: Michael Fenton
Development Editor: Cathy Sellars
Production Editor: Sirli Manitski
Marketing Specialist: Isobel Pritchard

Typeset by Kneath Associates, Swansea
Printed and bound in Great Britain by Bell and Bain, Glasgow

Contents

Foreword

In my career I have seen the huge range of roles and activities undertaken by civil engineers. While this variety provides great opportunities for diverse individuals to achieve personal fulfilment, it can be confusing.

This guide has been prepared by the Institution of Civil Engineers (ICE) to help those thinking about a career in civil engineering; to help them understand their options and how the different career streams complement each other. This is particularly important at this critical time in our planet's history as we need increasing numbers of skilled civil engineers to overcome the significant challenges that we face.

The guide is targeted at civil engineering undergraduates and those who have completed their studies and wish to map their career path.

The guide provides insights into the type of employers that there are within the sector as well as roles within those organisations. There are contributions from different types and sizes of organisations representative of the sector. The guide also showcases the different specialisms within the civil engineering discipline.

I believe that for many, reading the guide will be the first step on a journey towards a fantastic and rewarding career in civil engineering.

I hope too that the guide will mark the beginning of a long-term relationship with ICE, which will support you throughout your professional life. ICE is the oldest civil engineering professional body in the world and has over 96,000 members. It is a learning society and provides professional qualifications for civil engineers.

Through ICE you will have unparalleled access to a global civil engineering

community. You can take advantage of incredible networking opportunities with like-minded civil engineers to share and exchange best practice, keeping you and other engineers at the cutting edge of industry thinking.

Membership of ICE also unlocks access to a wealth of resources including the world's largest civil engineering library. These resources will help boost your ongoing development, keeping your skills and knowledge current, ready to meet the challenges of today and tomorrow.

Moreover, the collective wisdom and knowledge of our members helps ensure that ICE is the voice of the built environment, leading the infrastructure debate across the globe to create a more sustainable and inclusive future for all.

On behalf of ICE, I hope that you find the guide useful and encourage you to join our community of civil engineers as we transform lives through providing the infrastructure that we all need to survive and thrive in the 21st century.

Ed McCann, BEng (Hons) ACGI MSc CEng FICE FRSA FIESIS

Past President, Institution of Civil Engineers, 2021-2022

01 An Introduction to Civil Engineering

Introduction

Civil engineering is an exciting, rewarding, and diverse career, which offers the chance to work on a wide variety of projects all over the world, using your expertise and skills to help people and improve society.

There is a broad range of different ways that the work of civil engineers helps to support and improve society, for example:

Civil engineers design, build and maintain the infrastructure which supports daily life and that shapes the built environment – hospitals, schools, sports stadia, roads and harbours, railways and airports. They provide the infrastructure services that support our society and economy.

Equally important is the responsibility for adapting and maintaining the infrastructure that we depend on – our roads, railways and bridges; energy and water supply; waste networks and flood defences. Civil engineers keep this infrastructure running effectively and adapt it to meet challenges, such as population growth, climate change and natural disasters.

The infrastructure developed by civil engineers is vital for economic growth – providing better connectivity, job creation, access to services, and ensuring that the systems we rely on in daily life are resilient.

More widely, civil engineers also influence the development of society – advising governments, providing public health services including safe water, drainage and sanitation, delivering schools, pioneering new energy systems, planning the development of towns and cities, modernising transportation systems, and contributing expertise in developing nations.

Civil engineers are also specialists in the resolution of complex civil engineering challenges. These include addressing challenges of pollution, easing commuter congestion, supporting urban redevelopment, protecting communities from flooding, and combatting climate change. A career in civil engineering provides an opportunity to develop solutions to the problems facing society and to make a real and lasting impact to people's lives.

> "Civil engineering is the fundamental enabler of civilised life today and supporter of growing populations. It is an industry with an enormous range of opportunities."
>
> **Sir John Armitt CBE FREng CEng FICE**

Civil engineering covers a multitude of disciplines and activities and tends to attract a broad range of people because it draws on many different artistic and scientific talents. There are a wide range of opportunities and roles within the profession and an exciting and varied number of different pathways that a career in civil engineering can take. The purpose of this guide is to provide a key resource and planning tool for apprentices, undergraduates and graduates getting ready to embark on their career. It includes a broad overview of the different options that are available to provide a starting point in the job-hunting process, and it also provides an orientation on the industry, and insights across a range of sectors. The guide has been developed with the assistance of civil engineers at various stages in their career, and working across a broad range of sectors. It includes guidance on the range of options and connections between the different disciplines of civil engineering, examples of projects that illustrate the work of civil engineers, and individual career profiles providing insight into the day-to-day life of civil engineers. We would like to thank everyone who contributed to the development of the guide.

The guide is divided into three sections:

1. An introduction to ICE, ICE membership and the range of professional qualifications which can support career development

2. An overview of seven subdisciplines within civil engineering:

- Structures and buildings
- Geology, geotechnical and ground engineering
- Water engineering and wastewater management
- Coastal and maritime engineering
- Transport
- Energy and utilities
- Development, planning and urban engineering

This section includes key roles, case studies, and career profiles of engineers at different stages of their career

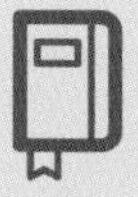

3. A list of contact details of all ICE Corporate Partners and ICE Approved Employers

Career pathways

Although there is a broad variety of opportunities and roles within each of the disciplines of civil engineering there are also some common career pathways that span the main disciplines:

Consulting and design:

Consultants get involved with planning and designing projects. Consulting engineers may be brought in at the start of a project to help with feasibility and costing, or may be called in later, when the client wants detailed design work carried out. Work includes preparing tenders, technical design, preparing design calculations, site surveys and detailed drawings.

Contracting and construction:

Contracting companies are responsible for constructing projects in a safe and timely manner, employing the labour, and bringing in equipment and materials to translate a client's ideas and the consultant's designs into reality.

Education:

Civil engineers in education are involved with teaching the next generation of engineers, and carrying out research aimed at developing advancements to the field of engineering. Graduate engineers seeking a career in education will often take an advanced degree, including a doctorate in many cases. Within many academic institutions, publishing research findings and obtaining grant funding for your institution are key elements to success.

Infrastructure owner / client:

The organisation or individual that commissions the project is called the client. Some clients do not have engineers on their staff full time so they bring in external consultants. But others – like National Highways, the Environment Agency, Network Rail, water companies and property developers – are responsible for building, running and managing assets that require full-time input from engineers.

Working for a client you could get involved with feasibility studies and outline design, detailed design, project management and, ultimately, managing the finished asset. Usually once a project has been planned by in-house engineers, the client hands it over to consultants and contractors to carry out the detailed design and building.

It is possible for a civil engineer to move between these types of employer. For example, some consultants will second staff to a contractor for work experience, or take in engineers who work on-site to provide them with high quality design experience. PhD graduates sometimes move into roles in industry which require specific expertise, and university lecturers may also provide professional consultancy on a part-time basis.

Industry profile

Civil engineering is a growing industry. An increasing population and the need for better and more secure infrastructure means civil engineers must find new ways of tackling these challenges. Governments around the world see infrastructure investment as key to re-booting economic growth, and this means that there's often a broad and interesting range of opportunities for graduate civil engineers starting out in their career.

Civil engineering employers range from small firms of less than ten people, to multinational companies with over 10,000. Civil engineering projects can also vary enormously in scale, from the construction and renovation of local buildings, to the construction of nationwide transportation systems, power stations, and venues and infrastructure for the Olympic games. The delivery of larger projects typically requires an extended supply chain providing a range of specialist skills, labour, materials, and services throughout construction.

> "We are the world's inventors, innovators and practical problem solvers. With every passing day, the challenge of climate change only grows greater. It is on us to step up and figure out what we can do to help address the problem."
>
> **Rachel Skinner CBE FREng CEng FICE**

Sustainability

A significant recent trend in the industry is an increased focus on sustainability and environmental protection. The climate emergency, loss of biodiversity, a global water crisis, and increased hazards from extreme events present the engineering profession with the greatest challenges it has ever faced. The individuals who are studying, training and starting out in the industry will have a vital role to play in addressing these issues. They will also be the leaders of tomorrow, with the opportunity to make a difference for current and future generations.

Civil engineers will play a vital role in addressing the climate crisis and the transition to net zero, both in developing innovative solutions that do not deplete the world's resources or contribute to carbon emissions, and in the provision of infrastructure solutions that actively encourage a shift to low-carbon lifestyles.

Solutions developed by civil engineers are also vital in providing social value and serving communities, including reducing inequality; maintaining natural capital and achieving net gains in biodiversity; management of ageing assets; and providing services in rapidly expanding future cities.

Many companies are now reporting on their alignment with the United Nations Sustainable Development Goals and identifying ways to accelerate efforts at reducing carbon emissions. Across sectors, civil engineering roles increasingly require individuals that are able to develop sustainable solutions and address a range of social, economic and environmental issues.

A career in civil engineering presents an opportunity to play a significant role in addressing critical sustainability challenges, and to work on projects that will have real and long-term benefits for people all over the world.

Digital transformation

A second industry trend is the emergence of digital transformation, digital engineering, and digital skills as key themes across the sector. Although digital technologies are not new to civil engineering, the landscape of digital technology is changing rapidly, and the pace of change is increasing. The development and adoption of digital technology is creating new techniques and processes that are changing the way that civil engineers work, and the outcomes that they are able to deliver. Digital technologies come in a variety of different forms. Some common examples include: building information modelling, artificial intelligence, 'big data' and robotics. Employers are increasingly looking for candidates who can develop an understanding of the efficiencies and other advantages that these technologies can bring, and who are able to apply these and other technologies to bring about better project outcomes.

> "ICE members all know that civil engineering is creative and exciting. The satisfaction of standing back and looking at what one has helped to create is indescribable. Civil engineering is a fascinating and rewarding career. It has and will continue to benefit from developments in science, technology, media and the arts."
>
> **Professor Lord Robert Mair CBE FREng FRS CEng FICE**

Key statistics

Qualified civil engineers are in demand across the construction sector, and exciting roles can be found working for consultancies, contracting organisations, local authorities, government departments, utility companies and environmental organisations in the UK and internationally. These are some key statistics about graduate careers:

Employment: A study by the Royal Academy of Engineering found 94% of engineering graduates in full-time work, pursuing further study or a combination of both three and a half years after graduation. This figure is 6% higher than for all graduates.

https://www.engc.org.uk/news/he-bulletin/november-2016/positive-employment-outcomes-for-engineering-graduates/

Graduate destinations: In 2019 Engineering UK found that most engineering and technology graduates go on to careers in engineering: 62% of full-time UK domiciled engineering and technology leavers entered an engineering occupation. Of all engineering disciplines, those studying civil engineering were most likely to do so (77.8%).

https://www.engineeringuk.com/media/156198/key-facts-figures-2019-final-20190627.pdf

Salary: The UK National Careers Service reports that the average starting salary for a civil engineer in 2022 is £30,000, rising to £70,000 for senior positions.

https://nationalcareers.service.gov.uk/job-profiles/civil-engineer

The Institution of Civil Engineers – your partner for a successful career

ICE is the world's leading civil engineering institution. We are committed to helping our members make a positive difference to society and to shape a better world. ICE is at the forefront of addressing the great challenges and opportunities we face, such as the drive to net zero and the integration of smart technology.

As you embark on a career in civil engineering, you can be confident that we will help you to build a successful, rewarding and transformative vocation.

ICE offers unrivalled support to help you to excel. Our team of membership experts, structured training programme and wider network of mentors will help

you to reach the high standards needed to become professionally qualified. Beyond this, ICE's cutting-edge knowledge resources and industry insight will help you to operate at the highest levels throughout your career.

The ICE vision

Our purpose has always been, and remains, to improve lives by ensuring the world has the engineering capacity and infrastructure systems it needs to allow our planet and those who live on it, to thrive.

Free ICE Student membership

ICE's free Student membership is the best possible start for your career, offering access to a host of benefits and invaluable resources to help you learn about the industry and improve your career prospects:

- Start the process to become professionally qualified. Take advantage of ICE's curriculum enrichment opportunities and record your experience
- Greater career prospects. Make connections and get valuable insight through industry journals and regular career talks
- ICE Rewards. Discounts at many different retailers, so you can make the most out of the time when you aren't studying
- Student & Graduate community. Meet loads of people with similar interests and take part in various informative or social activities
- Digital access to the world's largest civil engineering library, plus Ask Brunel for help answering any civil engineering question within 24 hours!

If you're a civil engineering student or apprentice, join ICE today to unlock access to these benefits and many more.

ice.org.uk/students

Advance your career with the mark of excellence

As you move beyond university and into professional life, we'll continue to give you all the support you need to develop as a civil engineer.

Our professional qualifications – Engineering Technician, Incorporated Engineer and Chartered Engineer – command respect. These are internationally recognised and will give you the skills and experience you need to progress your career and transform lives.

Achieving professional qualification will increase your professional status and influence with colleagues, employers and the public. 65% of civil engineers have post-nominals* and studies show it can open the door to earning up to £10,000 more per year.**

The Engineer Salary Survey 2021* and 2019**

Your platform to succeed

After finishing your studies, you can become an ICE Graduate member to accelerate your path to professional qualification. As a Graduate member you can access extensive support to realise your potential:

- Enrol on the ICE Training Scheme with an ICE Approved Employer, or follow our unique mentor-supported training to progress to professional qualification
- Get advice through our extensive online membership guidance resources and team of membership experts
- Record and track your progress with our market leading IPD Online system
- Demonstrate your status with 'GMICE' after your name.

Discover ICE Graduate membership:

ice.org.uk/graduates

Become part of a global civil engineering community

Around the world and at every stage of your career, ICE will be your partner for success.

Your ICE membership will open the door to our global community of over 96,000 built environment professionals in 150+ countries. The collective wisdom and knowledge of our members ensures that ICE is **the** voice of the built environment, leading the infrastructure debate across the globe to create a more sustainable and inclusive future for all. As well as our internationally recognised professional qualifications for civil engineers, welcoming Chartered Infrastructure Engineers will help shape the future of industry and the institution.

You can take advantage of unparalleled networking opportunities with like-minded civil engineers to share and exchange best practice, keeping you and other ICE members at the cutting edge of industry transformation.

Membership also unlocks access to a wealth of resources to boost your ongoing development – so rest assured that beyond achieving professional qualification you can continue to learn, keep your skills and knowledge fresh, and be ready to meet the challenges of today and tomorrow.

You can also access a range of financial and wellbeing support for ICE members and their families through the ICE Benevolent Fund.

Transforming lives

When you choose a career in civil engineering, you are choosing an exciting, rewarding and varied profession at a critical time.

Our world needs skilled civil engineers to overcome the significant challenges we face in the 21st century. As the world's foremost civil engineering body, we will help you rise to meet those challenges.

Civil engineers transform lives.

We'll help transform yours.

Join today at ice.org.uk/membership

How to become a professionally qualified civil engineer

Achieving professionally qualified status will unlock a world of opportunities for your career in civil engineering. You will gain a global professional passport that recognises your skills, knowledge and experience.

ICE's internationally recognised professional qualifications validate your abilities and give you a competitive advantage. 65% of civil engineers have post-nominals* and professionally qualified engineers earn an average of £10,000 more per year than those who are not.**

The Engineer Salary Survey 2021* and 2019**

ICE awards three professional qualification grades:

- Engineering Technician (EngTech MICE)
- Incorporated Engineer (IEng MICE)
- Chartered Engineer (CEng MICE)

To apply for these, you will need to have a minimum educational level and to have gained suitable professional experience. ICE accredits and approves academic courses so that what you study meets the requirements of the industry and also counts towards getting your professional qualification.

Whichever grade you are aiming for, we offer a range of options and extensive support to help you reach your goal.

How do you become professionally qualified?

For most people there are three stages, based on:

- Your academic qualifications (educational level)
- Your work experience (also called Initial Professional Development)
- Passing your professional review.

Academic qualifications

The educational level that you need can be met through accredited or approved qualifications, or through qualifications and/or professional experience that ICE has assessed as meeting that level.

For example:

- An approved level 3 diploma, approved HNC or HND or an accredited foundation degree are some of the qualifications that meet the educational level for **EngTech MICE** (there are many more)
- An accredited bachelor's degree meets the educational level for **IEng MICE**
- An accredited undergraduate master's degree meets the educational level for **CEng MICE**.

You could also have a combination of qualifications, for example a bachelor's degree and an MSc that meet the required level to become chartered, or you might not have any academic qualifications. Regardless of your educational level, there will be an ICE option for you to follow to the professional qualification you want to achieve.

You can find out more at **ice.org.uk/educationalbase**. You can check if your educational qualification is approved or accredited at **ice.org.uk/coursesearch**. If it isn't listed, contact us at aqp@ice.org.uk so we can check it out for you.

Work experience

Your **Initial Professional Development (IPD)** is where you develop the special skills, knowledge and experience that help you to become professionally qualified.

Your IPD is measured against a set of attributes – you can see more detail on these at **ice.org.uk/attributes**.

ICE offers a structured training programme to help you complete your IPD:

- **ICE Training Scheme** – run by your employer, approved by ICE. You'll receive support and guidance throughout your training from a supervising civil engineer (SCE), who your employer assigns to you
- **Mentor-supported training** – the ideal alternative if your employer does not offer the ICE Training Scheme. You're responsible for managing your own training with the support of an ICE approved mentor.

On either the ICE Training Scheme or mentor-supported training, you can record and track your progress with our market leading **IPD Online system** and our membership team will be on hand to monitor your progress in the workplace and offer expert advice at every step.

If you are aiming for EngTech MICE you don't have to enrol on these programmes, but they can help you to develop and achieve your goal. There are also other options for experienced engineers. Our membership team can advise which option is best for you.

Professional review

The professional review is the final stage in becoming professionally qualified.

This is where you prove that you've developed all the skills, knowledge and experience needed to become professionally qualified.

Having submitted an application, on the day of your professional review, you will be interviewed by experienced civil engineering professionals (your reviewers) and deliver a presentation. If you are applying for IEng or CEng MICE you will also complete a communication task.

Find out more about the key steps to professional qualification at:

ice.org.uk/professionalqualification

Supporting you every step of the way

ICE offers a unique and unparalleled level of support to help our members achieve professional qualification. This support is held in high regard by members and employers – for many it is why they choose to progress with ICE.

Whether you are following the ICE Training Scheme or ICE's mentor-supported training, these **structured development programmes** will set you on the path to success and ensure you develop the right skills to excel in your career. You will be supported by our **network of mentors** throughout.

Our **team of membership experts** offers advice and guidance on your path towards becoming professionally qualified, wherever you are in the world. They deliver a vast range of membership surgeries, workshops and courses both in person and online to help you every step of the way to professional qualification. They also monitor your progress working with you and your employer.

Our **extensive online membership guidance resources** will help you learn more about how to reach your goal through a comprehensive mix of bitesize videos, guidance documents and useful links.

More information

To explore our wealth of membership guidance resources, visit **ice.org.uk/membershipguidance**

For expert advice on your next step to professional qualification, contact our Membership Support Team at **membership@ice.org.uk**

02

Disciplines of Civil Engineering

Structures and buildings

Structures and buildings

Structural engineering is a specialist discipline of civil engineering dedicated to the design of structures in the built environment that can be constructed in a sustainable and safe manner. Structural engineers are highly skilled problem solvers that design the strength and stability of a wide range of structures to have robustness and the durability to resist external factors.

Structural engineers work alongside other construction professionals to create all kinds of structures from bridges, power plants, hospitals, airport terminals and skyscrapers to more specialist structures such as oil digging platforms and temporary structures.

It is common for structural engineering roles to specialise in the design of certain types of structure, some common areas include:

Bridges – design of bridgeworks tends to require sophisticated analysis and calculations, and interfacing with other disciplines, including highway and rail design, and geotechnical engineering. In bridgeworks, design and construction are closely linked – so bridge engineers need an understanding of both these aspects.

Tunnels – tunnelling is undertaken by various methods which affect the structural requirements. Structural engineering for tunnels is highly dependent on the ground conditions, as well as construction method. Structural design also tends to extend beyond the tunnel surround itself to include portal structures at each end.

Buildings – building structures vary hugely in scale and form. Structural engineering for buildings tends to sit within multi-disciplinary teams, with close interaction required with architects, mechanical and electrical engineers, planning permissions and construction methodology.

Underground structures – underground structures, such as underground stations or deep basements, combine different aspects of the requirements for structural engineering of ‘above ground’ buildings and tunnels, to which they are often connected. As with bridge design, construction methodology plays a huge part in the structural engineering of these types of structure.

Dams – with close interfaces between hydrology and geotechnics, structural engineering of dams includes design for huge forces, but also for associated structures required to service and operate the structure.

Offshore structures – this typically includes offshore wind turbines and platforms for oil and gas exploration and extraction. These structures are highly exposed and subject to substantial wind and wave loading. They are typically founded on the seabed requiring close interface with geotechnics and an understanding of the associated principles.

Some key skills needed to become a structural engineer include:

- Design skills and knowledge – an ability to solve structural engineering problems and produce structural design solutions using appropriate methods of analysis
- An ability to use your education, skills and judgement in a practical and pragmatic way, applying your theoretical knowledge to develop solutions to real-world problems
- It is also important to understand the impacts of a design, including the implications for health, safety, commercial, legal, environmental, social, energy conservation and sustainability
- Creativity and problem solving – a dedication to learn something new, think outside the box, and not being afraid to step outside your comfort zone
- Good communication and interpersonal skills are vital for working effectively with other professionals, both spoken and written; and also through sketching and digital channels
- Reliability in meeting deadlines, and attention to detail in making your way through checks on the implementation of a design, for example.

Professionals in this area often cite the following benefits of a career in structural engineering:

- The completion of a project often brings a real sense of achievement, in seeing something that you imagined and designed become part of the landscape and knowing that your skills were instrumental in bringing it into being.
- The projects also add real long-term value to society – creating safe buildings and infrastructure which is cost effective and sustainable.
- The work will often have a long-lasting impact – buildings are designed to last for 50 years, and bridges for over a 100. Structural engineers also breathe new life into old structures – renovating or changing the use of buildings that were designed decades ago and turning them to completely new purposes.
- It's an ever-evolving field where advancements in technology and materials mean you never stop learning.
- It also offers a range of different opportunities for career progression and to follow numerous different career pathways and specialisms.

Mae Ann Ta GMICE
Graduate Structural Engineer, AECOM

I enjoy being technically challenged

Working in an engineering consulting firm, my role as a structural engineer includes providing engineering services to companies in need of a specialised skillset for a build project. This typically involves providing structural advice, analysis and design solutions.

I enjoy being technically challenged on projects, by solving structural problems and creating solutions.

As a student I had the opportunity to design an office block during my placement year and was heavily involved during the early conceptual design stages. Although my involvement stopped just before work started on site, being able to see the project to completion and usage was a major highlight.

I am currently working on a project called Plymouth Bereavement Centre. This interests me because it has a relatively complex roof geometry, resulting in a design that was complicated but interesting to model.

Achieving net zero

Civil engineering is a rewarding profession. Our role as civil engineers is far more significant than in the past due to the climate emergency and developments that we are facing today.

I have seen a shift in civil engineering towards being more climate conscious. Civil engineering now prioritises sustainability in our work and professional development.

The profession places an emphasis on reducing embodied carbon, increasing climate resilience and adaptation, and enforcing mitigation strategies such as the reuse of existing structures (circular economy).

Achieving net zero carbon in any build or infrastructure project by 2050 will be a big challenge for the industry. Engineers will first have to face a design challenge which involves enforcing sustainable options through a collaborative approach with clients and architects. There will be a further challenge whereby low carbon materials (usage, research and development) will need to come into play to achieve the net zero target.

Helping me grow

I am working towards becoming a Chartered Engineer with ICE. The route towards professional qualification allows me to pick up skills that I never knew I needed at a personal and professional level. This process is helping me grow as an individual and an engineer.

As an ICE member, Initial Professional Development has been useful to me as it helps keep track of my progress towards a professional qualification, especially since coming back from my placement year. I also value the thorough, streamlined process and how it has been set up to be skill-relevant with the current times.

The skills that I am gaining on my way to chartership will keep me grounded as an engineer, and ensure my skills remain relevant to the ever-changing demands of industry and developments in technology.

Being professionally qualified will help me to be better at what I do and contribute more towards the engineering industry and society.

Career progression in structural engineering will often involve opportunities to get involved on larger projects and to take on more responsibility:

- Design Engineer – designing part or all of a structure as part of a consulting engineering practice. The level of involvement will depend on the size of the project and your level of experience.
- Senior/Project Engineer – leading a number of smaller projects or one larger one, managing client meetings, engagement with other consultants and delivery of technical requirements for the project.
- Associate/Associate Director – leading a team of engineers and technicians managing a number of projects and leading the technical delivery for a consulting engineering practice. Managing client meetings, engagement with other consultants.
- Director – leading a practice of consulting engineers, managing financial and commercial aspects including business development and work winning. Being the lead technical authority for a practice and leading the development of people within the company.

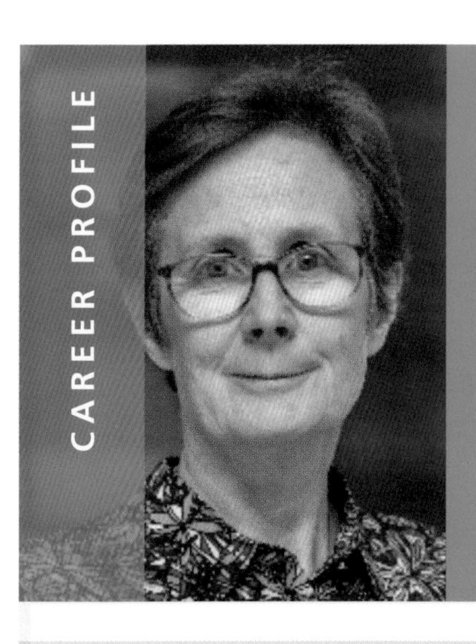

Dervilla Mitchell
CBE FREng CEng FICE
Deputy Chair, Arup Group

A role that is full of variety

The Deputy Chair role at Arup Group is full of variety and global in nature. I enjoy engaging with our people and clients from bringing them together and delivering successful projects.

My career highlight has to be Terminal 5 at Heathrow. It's the project on which I learnt the most and the collaborative environment was a huge enabler to all we achieved.

The most interesting project I have ever worked on was Portcullis House in Westminster – it is still the most technically challenging project of my career.

The most rewarding achievements in my career have been gaining Fellowship of a number of professional institutions and receiving honorary doctorates from University College Dublin and Imperial College London.

A golden age for engineers

Civil engineering is an interesting and varied career with lots of opportunities. I was inspired to become a civil engineer by my father who was an architect, and my grandfather and uncles as engineers.

My career has been really exciting and has exceeded all my expectations.

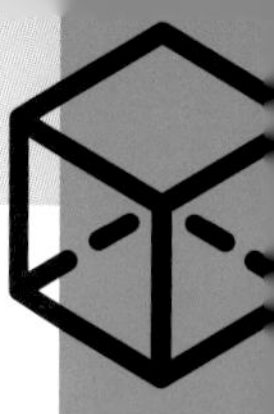

One great thing that I love about civil engineering is realising you can have an impact on people's lives.

Over my career the profession has transformed across many dimensions – digital, data and diversity to name just a few. The greatest challenge civil engineers will need to help overcome in the next 50 years I believe, will be resilience to climate events and other shocks. I hope we will increasingly shape people's lives for the better and bring a balance between humanity and nature in the work we do.

It's a golden age for engineers!

An important stepping stone

I was motivated to become a Chartered Engineer as it was the norm amongst colleagues and an expectation of my company.

Being professionally qualified is an important stepping stone in a civil engineering career.

To someone starting out in their career, I would recommend that they aim to get professionally qualified as quickly as they can (and don't forget to become a Fellow as well).

As a Fellow, I value being part of ICE's community. I would recommend ICE membership as it's part of your career journey.

> The greatest challenge civil engineers will need to help overcome in the next 50 years I believe, will be resilience to climate events and other shocks.
>
> Dervilla Mitchell

Structural building engineering

Structural building engineers ensure a building is structurally sound, taking into consideration load, material characteristics and stresses, ground conditions and climate variations. Using skills in maths and physics, sketching and computer modelling a structural engineer finds creative solutions to realise the architect's/designer's visions for a structure. A structural engineer will innovate, design and plan, collaborate and support a multi-disciplinary team towards a common goal – a performing and safe structure. Many organisations in this sector are now looking for structural engineers with expertise in sustainability and carbon reduction, and the concept of net zero carbon buildings is becoming increasingly important.

Specialties in this area include:

- **Tower engineering** – tall buildings are a common feature on the skyline of many cities throughout the world and present a number of unique challenges both in terms of design and construction – for example, the relative magnitude of lateral loadings to gravity loads increases significantly in taller buildings.
- **Earthquake engineering** – practitioners in this area focus on the seismic design of buildings, seeking to understand and analyse the way in which an earthquake stresses a building, choosing buildable structural forms and materials to cope with these stresses, and translating these concepts into practical, affordable and aesthetic spaces that people want to use and live in.

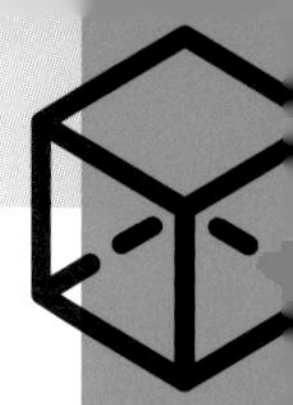

- **Façade engineering** – most large-scale commercial, industrial, educational and even residential buildings are now constructed using a frame and an envelope. Façades are one of the most complex, technically challenging and trans-disciplinary parts of a building. Specialists in this area consider the materials and components of the façade, thermal performance and analysis, weathertightness, fire performance and structural analysis.
- **Fire engineering** – specialists in structural fire engineering ensure that fire safety is built into the design of a structure wherever it is required.

CAREER PROFILE

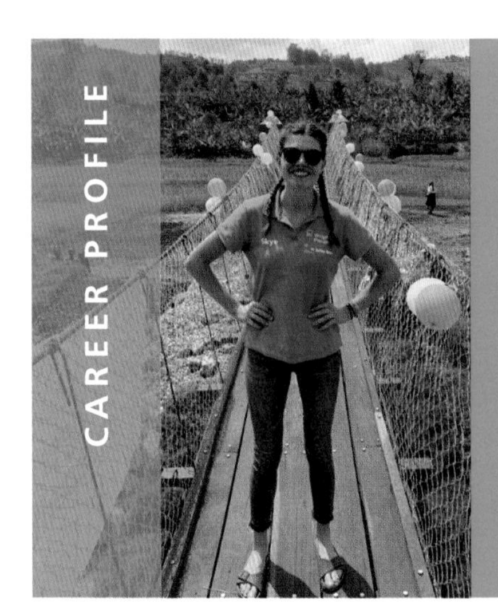

Skye Dick CEng MICE
Senior Structural Engineer, Mott MacDonald

A world of opportunities

I'm a structural engineer with a focus on lowering and reporting the associated embodied carbon of the projects I work on through my designs. I work in a buildings team, where one of the highlights of my career so far has been moving to New Zealand from Scotland with my company in 2020.

This move has been good for my career, as well as increasing my seismic design technical knowledge. In the last couple of years, I have become increasingly passionate and knowledgeable about the environmental impacts of structural design, including communicating the carbon associated with my designs to the client, contractor, and the community in which the project shall be a part of.

I particularly like the communication and co-ordination which comes with my job. As well as the opportunities to give something back to local and global communities. For example, I was part of a team of ten from Mott MacDonald and Balfour Beatty who built a bridge in Rwanda with the charity Bridges to Prosperity (B2P) in 2019. This was challenging and rewarding in equal measures. As well as assisting the local community to build the bridge, I also made lifelong friends with my teammates. This is what I like about engineering.

Contributing to net zero carbon

As an engineer, the carbon savings which can be made are significant compared to any personal differences that I can make. Over the next 27 years, civil engineers will help to overcome the challenge of building and maintaining the infrastructure required for an increasing population with no negative environmental effects.

I love that I can contribute to achieving the global net zero carbon goals through my work.

Fast-paced, diverse and ever changing

I particularly enjoy the fast-paced nature of my work. I work in an office with architects, electrical, hydraulic, service, and civil engineers. This involves a lot of co-ordination and talking through ideas and concepts. This is the best part.

When I was younger, I wanted a 'hands on' profession, although I wasn't sure what that was going to be. I was creative in school and interested in art and design, although I was also academic. My maths teacher suggested I looked into studying engineering – at the time I didn't even know what a civil engineer did. I went on a week-long course for 'Girls interested in Engineering', at the end of the week I knew I wanted to be involved in the design and maintenance of the built environment. I'd recommend anybody thinking of a career in engineering to look out for similar opportunities.

I would recommend a career in civil engineering because it is diverse, ever changing and requires 'big picture' thinking. The

profession needs a variety of engineers from all backgrounds and knowledge pools to work on global infrastructure challenges.

More professional responsibility

I'm a Chartered Engineer with ICE. I was motivated to become professionally qualified to allow me to further my career, demonstrate my experience and encourage others to do the same. Chartership is internationally recognised, and I am hoping this will help me move my career to other countries and opportunities in the future.

Being chartered allows me to take on more professional responsibility. It also allows me to mentor and assist younger team members as they develop their technical and soft skills – such as presenting to clients, communicating ideas, and bringing innovation to their tasks and projects.

> ...one of the highlights of my career so far has been moving to New Zealand from Scotland with my company...
>
> I would recommend a career in civil engineering because it is diverse, ever changing and requires 'big picture' thinking.
>
> Skye Dick

40 Leadenhall Street (Mace)

40 Leadenhall Street is a commercial office scheme delivered by Mace, in the heart of the City of London. The building offers 870,647 sq ft of office space and 28,860 sq ft of retail space and is one of the largest schemes ever to receive planning permission in the City of London. Rising up to 38 storeys, 40 Leadenhall Street is one of Mace's most digitally advanced projects, featuring a striking design and green credentials.

Mace's digital strategy for the project is based on the use of a model viewing and interrogation software allowing the team to digitally create all elements of the building before construction has begun. As a result, Mace has created a user-friendly 3D view of the asset information with all of its data visible for all project partners. The digital approach has made it easier for Mace to componentise and manufacture offsite the mechanical and electrical work on the project, saving time and avoiding clashes during module installation.

The delivery of 40 Leadenhall Street is connected to the cloud. Mace is able to track progress of material arrival times on site, tower crane hook times, labour on site and much more. Tracking and connecting the data helps Mace spot trends in the delivery of 40 Leadenhall Street, challenging its BIM capabilities and improving efficiency at every stage.

The structure comprises two large stability cores constructed using jump form methods, with the surrounding superstructure being a structural steel frame with structural metal decking floor plates. The substructure includes a 4-storey basement within a full secant pile cofferdam, constructed using a hybrid bottom-up/top-down approach, and is supported on over 600 large diameter rotary bored concrete piles. The cores are supported on large raft foundations, the largest single concrete pour being 1335 m^3.

In delivering the project, Mace has also introduced a sustainability strategy, slashing carbon emissions and reducing waste. Some of the innovative initiatives include the use of an electric concrete pump throughout construction, lowering the fuel usage of the concrete contractor by approximately 80%. Compared to using a traditional concrete pump, the carbon saving generated by Mace is estimated to be over 5 510 kg CO^2 in the space of four months.

Mace went a step further to introduce White D+ HVO as an alternative fuel type made from hydro treated vegetable oil. This replaced the use of white diesel, reducing emissions, NOx and particulate, improving air quality on site. The 1000 litres delivered to site thus far have reduced CO^2e by 2.8 metric tonnes in the delivery of 40 Leadenhall Street.

40 Leadenhall Street is one of the most complex projects that Mace is currently delivering, pushing the boundaries of data, digital and sustainability. The scheme is an example of how Mace leads the way to a more connected, resilient and sustainable world, working in partnership with clients and supply chain to create an iconic skyscraper by challenging convention.

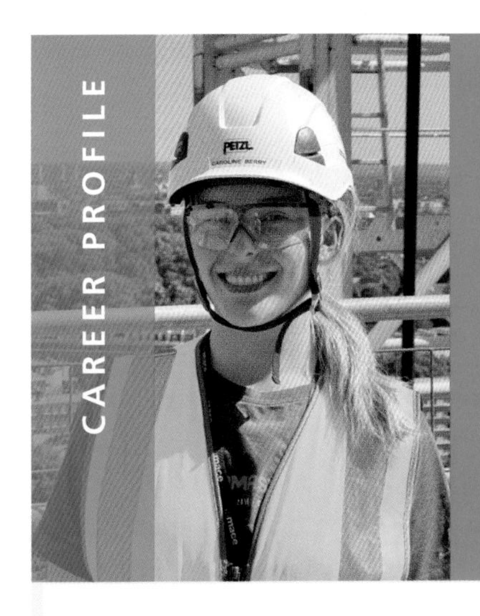

CAREER PROFILE

Caroline Berry

Assistant Engineering Manager, Paddington Square, Mace

What are some misconceptions you had about the industry?

Before I joined the industry, I assumed that everyone who worked in construction was a 'builder'. I didn't realise the vast variety of roles, from engineers and construction managers to quantity surveyors and planners – there really is something for everyone. I have also been pleasantly surprised at the diversity within my teams, and am working alongside a number of female engineers.

A project you wish you could have worked on and why

If I could have worked on any project, it would have been the Shard. It was the project that put Mace on the map for me. The range of construction methods used; from top-down construction to its hybrid concrete/steel frame, all while keeping London Bridge station in operation, is very impressive to any engineer.

If you from 3 years ago could see you now, what would they think?

Three years ago, I was 8 months into my degree apprenticeship with Mace. I was new to the industry and knew little of the engineering challenges that I would encounter on our projects.

Although I still have much to learn, I appreciate how far I have come. I would be impressed at how my confidence has grown, now routinely carrying out site inductions to up to 30 operatives and interacting with design engineers. I would be proud of how I have managed to balance work and university life, and I would be excited to see some of the projects I have worked on.

What does a typical working day involve?

In a typical day, I usually carry out inspections onsite; whether it's checking reinforcement has been installed as per the design before a concrete pour, or carrying out stud-weld inspections.

I am responsible for managing the site surveying contractor on the project – prioritising works, co-ordinating with other trades and analysing as-built surveys etc.

I also help close out NCRs (non conformance reports) for the structural packages as my current project is nearing completion.

> I didn't realise the vast variety of roles, from engineers and construction managers to quantity surveyors and planners – there really is something for everyone.
>
> Caroline Berry

What training have you received?

I am currently studying for a bachelor's degree in civil engineering. I have completed 4 out of 5 years alongside work commitments. Throughout my Apprenticeship at Mace, I have also received a lot of internal training. I have been on safety courses such as mental health awareness and first aid training, technical course such as temporary works co-ordination, SMSTS and concrete appreciation and courses to develop soft skills such as presentation skills and personal branding. It is really important to have a range of skills to become a well-rounded engineer who is adaptable. Mace also support my professional development and I will soon be taking my Incorporated Professional Review.

The best pieces of your job

I love the range of people I get to interact with on a day-to-day basis; from operatives, the client, architects, and third-party members such as the neighbours. I also enjoy problem solving and getting to see the progress of a project.

How has your career progressed from the beginning to now?

Over the past 4 years I have gained a lot more responsibility – managing small subcontract packages and signing off pre-pour inspections. I am currently working on my third project meaning I have formed some great connections throughout my career thus far.

Bridge engineering

A bridge engineer is a civil engineer who specialises in the design, construction, repair and maintenance of bridges. All bridges require optimal designs to ensure resistance against wind, tides, earthquakes and other external forces, and developments in materials, equipment, and construction methods have enabled bridge engineers to develop structures that cross greater distances in safer ways, while minimising environmental impacts. Some roles focus on the design of new structures, while others concentrate on inspection and planning the rehabilitation of aging bridge infrastructure. Engineers will prepare plans, specifications and cost estimates and, during planning and design, consider what type of bridge will best meet the needs of the area and how the selected building site will support it. Bridge engineers also perform load rating and stress analysis calculations to ensure that the structure can stand up to the weight it will bear and the environmental stresses of the site. They must also take into account local and regional building codes and requirements. Those in project management roles are tasked with overseeing bridge construction and ensuring that projects stay on schedule and on budget. They are often on hand at building sites to manage and monitor building progress. They ensure that all aspects of bridge development run smoothly, and resolve any problems encountered during construction. These engineers work with construction crews, subcontractors, suppliers and others to ensure that construction is completed safely and successfully.

200 years of bridge engineering

◀ Menai Suspension Bridge (1826). 176 m main span. The world's first major suspension bridge and still going strong 200 years later! Thomas Telford's innovative thinking and application of structural engineering principles connected Anglesey to the mainland through a reliable and safe method, avoiding the need to transverse the dangerous straits below. A brilliant structure, and life-changing for those living nearby.

▼ Golden Gate Bridge (1937). 1 280 m main span. One of the world's most iconic structures. Built 100 years after Menai with the same concept that Telford developed. But using advances in material engineering developed in between to span further, and in a region subject to regular earthquakes. Again, a great example of good design using the latest engineering science to improve people's lives and produce an elegant and beautiful bridge.

1915 Çanakkale Bridge (2022). 2 023 m main span. The same challenge and the same concept as Menai and the Golden Gate. But using the latest in engineering science, analysis and construction methods to elegantly span further than ever before.

Øresund Bridge

Build a crossing to connect the Danish capital Copenhagen and the city of Malmö in Sweden

The Øresund Bridge is part of the Øresund fixed link - a bridge and tube tunnel route connecting the Danish capital Copenhagen with the city of Malmö in Sweden.

The link is made up of an 8 km-long bridge, a 4 km underwater tube tunnel and 4 km of a reclaimed island in Danish waters called Peberholm. The bridge part of the link is the longest rail and road bridge in Europe.

The bridge carries rail and road traffic on a dual-track railway and a four-lane highway. Project designers chose a tunnel for part of the crossing amid concerns that a bridge in that part of the strait could interfere with radio signals from nearby Copenhagen Airport.

Other reasons for a tunnel included providing a clear channel for shipping in all weathers and the need to prevent ice floes from blocking the strait.

One of Scandinavia's largest ever investments in infrastructure, the link is jointly owned by Sweden and Denmark.

The crossing opened on 1 July 2000, with a ceremony hosted by Queen Margrethe II of Denmark and King Carl XVI Gustaf of Sweden.

Difference the project has made

The Øresund Bridge and fixed link have connected two major metropolitan areas – Copenhagen, the capital of Denmark, and the Swedish city of Malmö.

The bridge is credited with helping to create a trading region of 3.7 m people. It's certainly made it easier for goods and people to move between Sweden and Denmark – a one-hour ferry trip was replaced by a 10-minute high-speed rail journey.

The bridge carries a data cable as well as road and rail traffic. The cable is viewed as one of the most important routes for data transmission between central Europe and both Sweden and Finland.

How the work was done

The 8 km-long bridge section of the Øresund fixed link is the world's longest cable-stayed bridge for combined road and high-speed rail traffic.

The bridge crosses the shipping channel of the Øresund Strait in a 490 m-long curving span. Engineers designed the bridge with two side spans of 160 m and 141 m – one on each side of the main span.

The project team used a composite of steel and concrete for the bridge's main supporting girder. The concrete top deck of the bridge carries road traffic. The lower deck carries two railway lines.

Each of the bridge's concrete pylons are 203.5 m high and founded on limestone.

Most of the components of the bridge – such as the caissons and piers – were prefabricated on-shore and then floated out to the construction site for assembly.

A caisson is a watertight structure used to help construct parts of a bridge. Floated into place and lowered to the seabed, they provide engineers with a dry environment to work in. A pier is an upright structure that helps support a bridge.

Engineers built artificial islands around and near the bridge to protect against ships hitting the structure.

Gull Wing Bridge, Lowestoft (Farrans Construction)

Farrans Construction is on site constructing the Gull Wing, Lowestoft's third crossing over Lake Lothing, for Suffolk County Council.

Construction work began in spring 2021, with trial excavations taking place to confirm locations of utilities such as water and gas pipes.

The bridge opening is anticipated to take place in the summer of 2023.

The North Approach Viaduct (NAV1) section arrived at the construction site on Tuesday 1 March 2022. The major steel section of the bridge was transported on a barge following a 32-hour crossing from Ghent in Belgium.

Fabricated by Victor Buyck Steel Construction, NAV1 is approximately 55 m long.

On delivery it weighed 380 tonnes and, when ready to move into its final position spanning the East Suffolk railway line, it will weigh approximately 1 450 tonnes.

The complex process of navigating the barge into the inner harbour and then transferring the steel structure from the barge on to the site was overseen by a team of experienced civil engineers, port masters and marine experts.

It was then moved to a special platform where a concrete slab deck was cast in situ before the entire span was moved and placed in its final position during a weekend-long railway possession.

NAV1's arrival was dependent on weather conditions remaining favourable and it needed to arrive during a slack tide to travel into the inner harbour.

Neil Barnes, regional director for Farrans, said: "The arrival of NAV1 was an important milestone for the progress of the Gull Wing project and also for our team, who have been working hard to prepare for the technical challenges involved in receiving and moving such a large section of the bridge.

"Victor Buyck began manufacturing this structure back in July 2021 and we are pleased to now have this integral piece on site, following meticulous organisation in both Belgium and in Lowestoft".

The Gull Wing Bridge will reduce congestion and improve connectivity, something that is central to the regeneration of not just Lowestoft but the wider Suffolk and Norfolk region.

During construction of the Gull Wing Bridge, the project will offer in excess of 50 employment and training opportunities for the local area.

Farrans has significant experience with large scale infrastructure projects most recently including: the West Cumbria water supply pipeline, Edinburgh trams to Newhaven, M80 Steps to Haggs and Northern Spire Bridge. We have successfully completed in excess of 250 projects in the region over the past three decades through our building and civil engineering divisions.

Watch our recent project video: youtube.com/watch?v=JOcH1wkGCxs

Next steps...

Structural engineers are currently in great demand due to playing an integral part in major projects such as HS2 and Thames Tideway. There's a wide variety of employment opportunities, and graduate engineers work across a range of industrial sectors, including construction, aerospace, automotive, renewable energy, and oil and gas.

Engineering graduate schemes offer a real-world, hands-on introduction to the industry, and opportunities to gain experience across a range of different types of project. Relevant work experience through internships and industry placements may give you an advantage in obtaining a place. As your career develops you will have the option to specialise in particular areas of work – for example, in conservation projects, sustainable building materials or forensics, where you investigate why a building or structure has failed. There are also opportunities to move into project management; research and lecturing; freelance consultancy work; and on construction and engineering projects overseas in international development and with disaster relief agencies.

02 Disciplines of Civil Engineering

Geology, geotechnical and ground engineering

Geology, geotechnical and ground engineering

Geotechnical engineers are vital members of diverse multi-disciplinary teams, supporting the delivery of projects across a wide variety of sectors, including flood protection, dam designs and waste containment technologies.

Geotechnical engineers play a crucial role in planning, design and delivery of infrastructure projects by providing a detailed understanding of the structure and behaviour of ground conditions, and in the design and construction of structures in the ground. The discipline covers all aspects of engineering projects that are to do with support from, or support to, the ground. The work of geotechnical engineers is essential for ensuring that our structures and infrastructure are safe, don't collapse and don't move more than the design allows.

Geotechnical engineering is a broad and varied branch of engineering, and an aspect of most projects. It covers desk studies that consider information on geology and site history, ground investigation and characterisation of the ground, through to geotechnical design and construction. Careers in geotechnical engineering provide the opportunity to work across a wide variety of different sectors, including buildings, infrastructure for water and energy, railways, highways, airports, port developments, offshore structures and mining.

Although the structures that geotechnical engineers are responsible for are often not visible to the public, ground engineering is nonetheless an area that touches almost every aspect of modern life:

Foundations for buildings and structures: schools, hospitals, bridges, hydro-electric dams, wind farms and nuclear power stations, offices, sports stadiums, shopping centres, and homes.

Earthworks for essential infrastructure: roads, railways, airports and flood defence embankments.

Environmental protection: clean up of contaminated land, landfill design, groundwater management.

Hazard control: assessing and managing the stability of cliffs and slopes; stabilising underground voids, including natural cavities and old mines.

Underground structures: tunnels, metro systems, car parks, water and sewerage transmission tunnels.

Engineering for the future: ground source heat pumps, wind farms, protection against rising sea levels and coastal erosion.

(Courtesy: The Ground Forum)

Two main areas of specialisation within geotechnical engineering are:

1. **Ground investigation** – this is the process by which ground engineering professionals, such as geotechnical engineers, engineering geologists, and geoenvironmental engineers and scientists, obtain information on the ground beneath sites that are to be developed. There are many techniques, each appropriate to different ground conditions and engineering properties, but some of the key elements are:

- desk studies to review a range of different sources of information, including geological maps and aerial photographs from different time periods.
- ground investigation, including the drilling of boreholes and test pits to collect soil and other samples. Ground investigation supervisors record drilling activities (rock and soil logging), and supervise drilling and site investigation, analysis of soil and rock samples and groundwater.
- laboratory testing to evaluate the engineering properties of soils and rocks.
- development of factual and interpretive ground models which are used to establish the geotechnical parameters for use in design.

Related areas include:

- **Engineering geology** – engineering geologists use their knowledge of geology to understand how the geological past has impacted the way the ground will behave in response to both natural processes and construction works.
- **Material science** – the branches of materials science which ascribe engineering properties to soil and rock are known as soil mechanics and rock mechanics, respectively. The behaviour of real soils is complex and we have to idealise it through models of soil behaviour. Due to the inherent variability in natural ground, experience and judgement are also required in determining appropriate soil parameters.

2. **Geotechnical design and construction** – geotechnical design plays a key role in most civil engineering projects. Geotechnical designers need to utilise the conditions found on site to design and supervise the construction of a range of geotechnical structures, including foundations, slopes, excavations, tunnels, and retaining walls. Career development in this area will typically begin with simple engineered solutions like shallow foundations and retaining walls, and with experience it will develop to deep foundations, geogrid design and ground improvement using finite element analysis.

Some common areas of work include:

- **Underground structures** – geotechnical engineers play a key role in the design and construction of tunnels, subways and other underground facilities that are used in tunnels, as well as underground transport links and railways, waterways and waste storage. Geotechnical engineers are often present throughout the construction process to support the development as it progresses and faces new challenges.
- **Deep foundations** – all buildings, including high-rise structures, bridges, towers, and ports have a foundation that is carefully designed by teams of geotechnical engineers to ensure it can withstand the surrounding environmental demands.
- **Deep excavations** – geotechnical engineers ensure the stability of deep excavations – shoring operations allow deep excavation to take place in urban areas, which are vital for the construction of underground railways, roads, drainage and more in heavily built-up areas.
- **Transport and infrastructure** – geotechnical engineers ensure that important infrastructure such as roads, highways and railways are properly designed and maintained to ensure their longevity. This can also include measures required to protect roads from landslides, and foundations for bridges.

- **Offshore structures** – geotechnical engineers design the foundations of offshore wind turbines, and structures for oil and gas production.
- **Landslides** – assessment of surrounding environments is important to determine the risk of landslides that threaten public safety. Geotechnical engineers provide reports on the likelihood of risks and design the corrective measures to ensure public safety.
- **Scour and erosion** – natural processes such as longshore drift and extreme weather can erode the natural landscape. Heavy rainfall, snowfall, hurricanes and floods can rapidly change the landscape; therefore, in some cases, it is necessary to mitigate against this damage if construction work is due to take place or it poses a danger to life.

- **Dams** – these are large-scale geotechnical engineering projects which often involve assessment of the bed of lakes and rivers, the examination of shorelines and the wider effects to the surrounding ecosystem that a dam will have.
- **Temporary works** – temporary works are a critical part of civil engineering projects, enabling the permanent works to be built. Geotechnical engineers are involved in the design of temporary works which support the ground during construction.
- **Landfills** – waste disposal sites must be carefully planned to ensure that solid waste is isolated from the environment to protect public health. Landfills contain industrial, agricultural and household waste that sometimes contains harmful chemicals, pesticides and biological compounds. Specially designed systems are required for separating these elements, preventing harm to the public and the environment.
- **Contamination** – contaminated land can pose a threat to public health and the environment, and geotechnical engineers are often involved in making assessments and determining the best course of action. Geoenvironmental engineers use numerous techniques to remove contaminated soil.
- **Ground improvement schemes** – geotechnical engineers use a range of ground improvement techniques on poor or unsuitable subsurface soils so that they can support civil infrastructure that would otherwise not be possible. For example, the importing of soil and material to improve the stability and safety of the ground surface.
- **Mining** – mining and drilling excavations deep below ground pose significant risks to both the health of employees and the surrounding environment, so a thorough assessment is necessary to minimise danger.
- **Ground water** – hydrogeologists analyse and design groundwater flow and control – this field encompasses the exploration and development of geothermal energy resources.

CAREER PROFILE

Katie Askew

Site Engineer, JN Bentley

How would you describe your current role? What do you enjoy about it?

Currently I am a site engineer, working on mine water treatment schemes for the Coal Authority up in Scotland. This involves installing a lot of new pipework, concrete structures, and earthworks – so a really varied project! The variation is what I enjoy about the job as we work our way through different elements of works, each day brings new challenges and being able to see the projects develop and take shape creates real job satisfaction. Being on site also allows you to work alongside other teams, such as geologists and archaeologists so it is interesting to see how they work.

What stands out as the most interesting project you've worked on?

All projects I have worked on have had really interesting elements, and all have been really different so far. My first site project involved the modification of a historic pumping station, where we had to install and connect a ring beam to the existing below ground structure while demolishing the above ground to create safer access and operation. Unfortunately, I never got to see the finished job as I had to go back to university!

Career development, education, first role, career highlights and projects

I went to Northumbria University where I gained a BEng in civil engineering, after my geography school teacher suggested engineering as a path. While at university, I completed a year in industry with JN Bentley where my time was split between the design and site teams, so I was able to get a real insight into the whole process, from assisting with bids for projects to installing flood defences. When I came back to JN Bentley after graduating, I was able to work on the project I had been bidding for a few years previously. Since returning, I have continued to enjoy a mix of design and site work, which has vastly enhanced my engineering knowledge – and I am still less than 2 years into my career! My highlight is the most recent project I completed: we had to install a manhole chamber over an existing pipe – which should have been simple but unfortunately the condition of the existing pipe was horrific. It was a really challenging process and build to be a part of, but the end result looked amazing.

“I have continued to enjoy a mix of design and site work, which has vastly enhanced my engineering knowledge – and I am still less than 2 years into my career.”

Katie Askew

What changes have you have seen and where do you think the profession is heading?

In the short time that I have been working on site, the advancement in technology that we are being encouraged to use and trial is really increasing. Being able to get GPS attachments for machines really allows high level accuracy and control – which is particularly beneficial on larger scale projects as well as reducing the risks of us as engineers getting in and out of excavations. This therefore allows us to deliver great quality projects to our clients. I'm excited to trial more new technology as it becomes available, and to see how it can further improve the quality of work we are able to deliver.

> each day brings new challenges and being able to see the projects develop and take shape creates real job satisfaction.
>
> Katie Askew

Tunnel engineering

There are many reasons why tunnels or other underground excavations are required, and many methods for their construction. What they all have in common is the need to provide a conduit or space under or through an obstacle, be it a mass transit system under a busy city centre; a high-speed rail line underneath a mountain range or the sea; a road link underneath a river; an oil, gas or electricity pipeline; or a water supply or sewer tunnel for a city. The method employed for the construction of a tunnel depends on the length and size, but most importantly on the ground and groundwater conditions through which the tunnel is built.

There is a range of different methods of tunnel construction, each requiring specialist skills and equipment. These include techniques such as bored tunnels using tunnel boring machines, sprayed concrete lined tunnels, tunnels constructed by blasting with explosives, and tunnels constructed by pipejacking, piperamming, or thrustboring. This large variety means the tunnelling world and the skills it needs are constantly varied and challenging.

Bringing value to society

I am a geotechnical engineer working on large-scale infrastructure projects for Rambøll Denmark.

My role focuses on multi-disciplinary project coordination mixed with some technical design. I enjoy bringing people together while delivering value to the client. Whenever possible, I communicate our design externally – engagement with local communities is vital to me.

I have focused my career on infrastructure projects to utilise my skills in bringing direct value to society. I worked on deep basement designs in London, the foundations of one of the largest bridges in Denmark, and a train station design of one of the busiest hubs in Copenhagen.

There are small recreational canals in Copenhagen that I am particularly proud of. They meander among residential buildings and a newly built Metro station to connect to the sea and create a Venice-like feeling for residents. Improving people's well-being and opening the city to green areas was a great success and a rewarding feeling.

A career that can take you places

I usually work 2 – 3 days from home, and the remaining time at the Copenhagen office. I also visit construction sites or client

offices. My job is fully flexible, so I am able to mix my job and travel plans and work from other Rambøll offices around Europe (Oslo is my favourite!).

Civil engineering can take you places. Moving across the globe, travelling, and exploring new cultures and customs while contributing to society is particularly interesting. Plus, seeing your projects come to life and being used by engaged communities is the most rewarding feeling you could imagine.

Being a member of ICE has widened my social and professional network. I have many friends and colleagues working on projects all over the world. I can always reach out for help or advice, which is useful when working on many international projects.

Industry challenges and evolution

Our planet is facing unprecedented pressure due to climate change. As civil engineers, we should be frontrunners of the sustainability agenda and role models to other industries to help overcome this challenge. We have the necessary skills to answer some of the challenges ahead of us but will need to develop new ones for new problems emerging with time. The combination of unpredictability, innovation, and being close to the heart of many solutions is particularly exciting in my job.

One of the greatest shifts that is happening right now is the digitalisation of our industry. Civil engineering has embraced coding, parametric design, paperless offices, digital site visit reports, health and safety reporting apps... and more.

Having in mind how disruptive this change is, I cannot wait to see what the new generation of civil engineers will bring to the industry and our institution. I would love to see more AI and machine learning applications. No matter what, the ability to learn and embrace new technologies will be a vital skill for any newcomers and existing practitioners.

Some key skills needed to be a geotechnical engineer include:

- **An interest in problem solving and how things work** – ground engineering normally involves examining the detail to understand the bigger picture. You'll need to apply technical knowledge to problems and gain a deep understanding of the project as there will be no perfect textbook answer. This requires making judgements and decisions, which sometimes involves researching topics independently. With this comes someone who is confident in presenting and discussing ideas, and also someone who is a good listener, engaging with other team members and listening to their ideas.
- **Communication skills** – geotechnical engineers are involved in most civil engineering projects and are normally some of the first involved, so it's important to be able to communicate and interface with other disciplines. Eventually, you will be managing teams of people so this is key.
- **Keeping on top of a constantly changing field** – geotechnical engineers often have to deal with new equipment, updated safety regulations and other

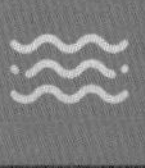

factors – above and beyond their duties to a client's construction project. With this in mind, keeping on top of scientific and industry news is an important habit.

- **Innovation** – it's also important to be innovative in this discipline, building a greener future for the growing population and factoring in other major social, economic, and environmental factors that will be a catalyst for change.
- **Report writing** – dependent on the role, you will have to communicate reviews of site information and conditions, ideas, design solutions and calculation summaries to the client, and also internally to your team. This is important as ground risk is often a significant risk in engineering projects, attracting significant sums if not adequately managed. In addition to report writing, at a later stage in your career the ability to manage finances and discuss commercial decisions is also important.
- **Digital skills** – these are increasingly of benefit in geotechnical engineering and involves being able to model ground conditions, simulating construction and examining the ground movements and soil–structure interaction due to construction activities.

PROJECTS

The Channel Tunnel

The Channel Tunnel is a 50.46 kilometre railway tunnel that connects Folkestone with Coquelles beneath the English Channel at the Strait of Dover. It has the longest undersea portion of any tunnel in the world (37.8 kilometres), and has been recognised as one of the 'Seven Wonders of the Modern World'.

The average depth of the tunnel is 50 metres below the seabed with the lowest point 75 metres below. Much of the chalk marl spoil bored on the

English side was deposited at Lower Shakespeare Cliff in Kent, now home to the Samphire Hoe Country Park.

Eleven boring machines were used to dig the tunnel. Together they weighed a total of 12,000 tonnes (more than the Eiffel Tower) and each was as long as two football pitches. Since opening in 1994, more than 465 million passengers have travelled through the Channel Tunnel – the equivalent of three times the populations of the United Kingdom and France combined, and 450 million tonnes of goods have been carried through the tunnel. Trains travel through the tunnel at up to 169 km per hour, taking 20 minutes to get from one end to the other.

Geotechnical engineers often highlight the following benefits of a career in this area:

- Ground engineering allows you to be creative, to work towards and develop solutions that will be different on every job you work on – maybe even different across a site – and you will quickly develop skills in managing people and finances.
- One of the most enjoyable parts of geotechnical engineering is the excitement of facing the unknowns! Every project offers its own challenges, and every site is underlain with a different ground condition and its history. It's an everyday challenge to predict the ground response and design the best foundation solution that would fit for the purpose of its use.
- There is tremendous variety in this sector and an opportunity to work across a broad spectrum of projects. There are opportunities both in office roles and on site, and even in the office there is a usually a good balance with site work and site visits.
- Many professionals in this area enjoy the collaborative problem solving that's involved in geotechnical projects – engineers work in teams, sharing ideas and experience in order to move forwards with efficient and effective solutions.

Next steps...

Graduate schemes in ground engineering are available at most civil engineering firms. There is a growing preference from employers for an MSc/MEng in geotechnical engineering or in related disciplines such as soil mechanics, rock mechanics, engineering geology, geophysics or hydrogeology. Relevant vacation work or an industry placement can improve your chances and it is great if you can demonstrate through your MEng that you have a keen interest in geotechnical engineering; i.e. through your dissertation, project work or engagement with the course. Alternatively, some companies offer a degree apprenticeship scheme in geotechnical engineering.

Graduate opportunities exist across a broad range of roles in ground engineering, including engineering geology; with ground investigation companies and geotechnical laboratories; in civil engineering design firms; and in contractors in both construction and temporary works design and construction. In addition to roles within civil engineering design firms, there are also openings within mining and oil and gas industries; with firms in the energy industry that are working on sustainable energy sources, and opportunities relating to foundation design in particular.

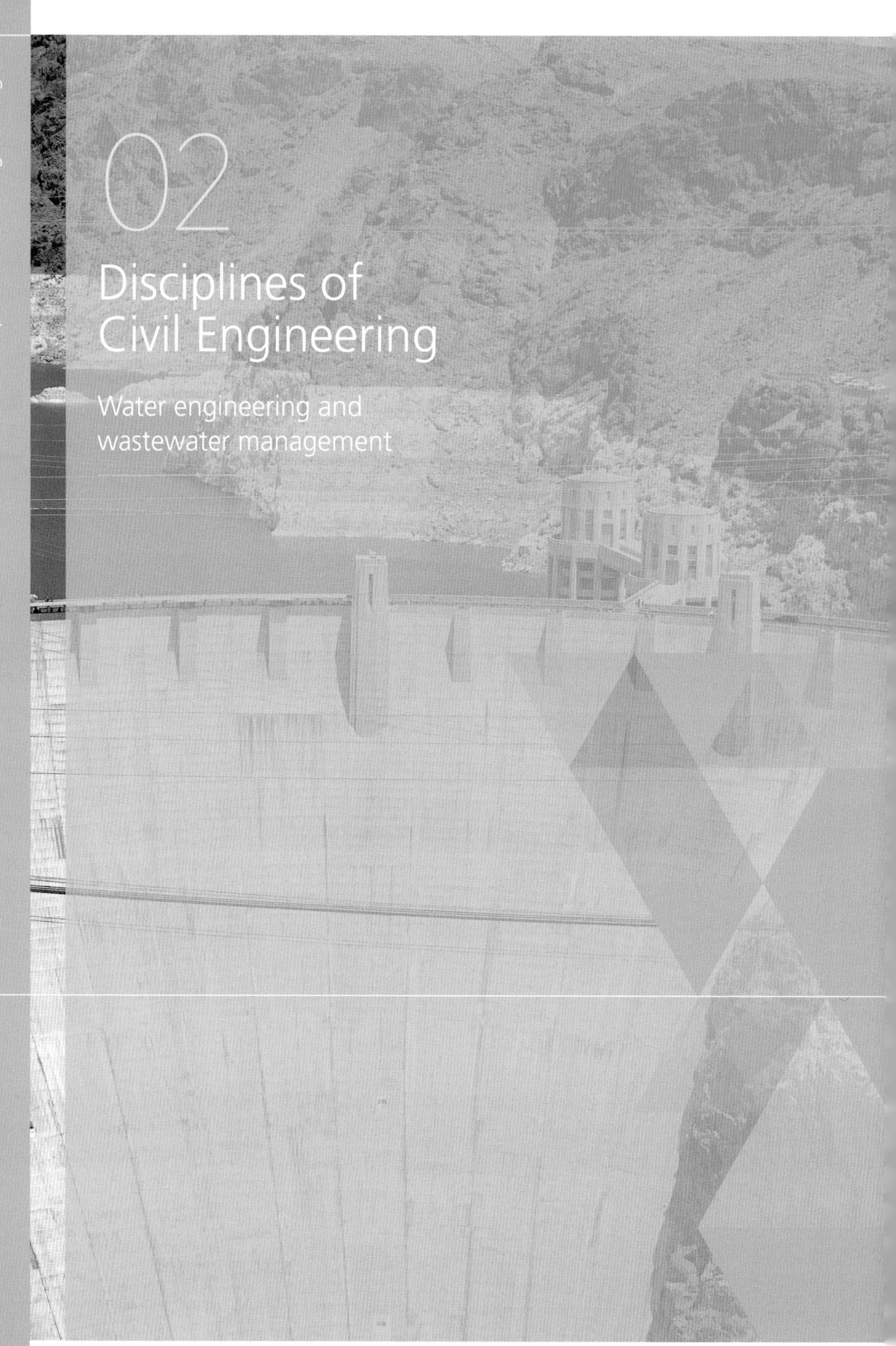

02 Disciplines of Civil Engineering

Water engineering and wastewater management

Water engineering and wastewater management

The successful design, development and maintenance of our water and wastewater infrastructure requires the integration of many disciplines. Careers in the water sector span the full spectrum of water and wastewater management issues in the natural and built environment, including water treatment, supply and recycling, and management of flood risk.

Civil engineers in the water sector provide the essential infrastructure which ensures that the public receive clean drinking water, have wastewater removed and treated, and ensures that rivers are managed and monitored. Work in this sector involves design, construction, maintenance and repair of structures that control water resources, such as dams, reservoirs, and pipelines to provide safe water;

sewers, sewage treatment works, sewage outfalls to prevent pollution and disease; and irrigation and drainage schemes to increase arable land.

Some key areas of work include:

- **Water resources management**
- **Water treatment and supply**
- **Wastewater infrastructure and treatment**
- **Dams and reservoirs engineering**
- **River and waterway engineering, including natural flood management and river restoration**
- **Water environmental management**
- **Flood and coastal risk management (see also maritime and coastal engineering)**
- **Sustainable drainage systems and green infrastructure**
- **Asset management**
- **Climate change adaptation and resilience**
- **Catchment science and management**

Roles in this sector often involve:

- Undertaking feasibility studies and preparing technical reports
- Designing and building water storage facilities such as dams and reservoirs to ensure the continued delivery of water during drought periods
- Producing designs of water treatment, sewerage, and flood defence structures such as pump systems and pipe networks
- Specifying and interrogating survey information, hydraulic and coastal risk modelling
- Calculating the future water demands for an area, based on population data, usage and storage
- Assessing water quality implications and developing measures for drought management
- Developing and appraising options and supporting the production of business cases
- Meeting with clients, local authorities, and other stakeholders
- Enforcing health and safety regulations and water legislation
- Forecasting and managing budgets
- Keeping project deliverables on track.

Some key skills to work in this sector:

- Graduates are often encouraged to develop an understanding of the many different elements within the water industry, such as civil, hydraulic, structural, geotechnical and environmental design through calculations, drawings, report writing and use of engineering software
- Some projects may also require an appreciation of other scientific and engineering disciplines such as electrical, mechanical, instrumentation, biology and chemistry
- Effective communication and collaboration is a key skill, as you'll be working with a wide range of organisations, people and stakeholders including members of the public, so it's important to take everyone's opinions and thoughts on board
- Attention to detail and accuracy is needed – when laying down a pipe or creating a treatment plant, for example, it is important to avoid clashing with existing utilities
- Digital skills are becoming ever more important, for example in data science, artificial intelligence and machine learning.

The development of London's sewer system

In the mid-19th century, London was suffering from recurring epidemics of cholera. More than 10,000 Londoners were killed by the disease in 1853–1854. The hot summer of 1858 created the 'Great Stink of London'. This, together with the frequent occurrence of cholera, gave impetus to legislation enabling the metropolitan board to begin work on sewers and street improvements. Civil engineer Joseph Bazalgette, who had just started as the chief engineer of London's Metropolitan Board of Works, was put in charge of the project. Bazalgette's solution to the city's health problems was to build an extensive underground sewer network that diverted London's waste downstream to the Thames Estuary – away from the main areas where people lived. Bazalgette spent 9 years digging up London to create six 'interceptor' sewers, which were around 100 miles long altogether. Another 450 miles of sewer fed into them. It was an incredible feat of Victorian engineering, and the sewers made the single greatest contribution to improving the health of Victorian Londoners. The development went beyond the immediate goals of the parliamentary vote that enacted it, and led to another step-change in urban health, economic confidence and industry. The bulk of the system remains in use today.

Bazalgette: Saviour of the Great Stink:
youtube.com/watch?v=5k8AnhNkN04

Fast-forward 150 years, and although the system still works well, it's now struggling to cope in terms of capacity. London's sewage system was originally designed for a city of around 4 million people, there are now around 8 million people living in the capital. The result is that millions of tonnes of sewage pour into the river Thames every year.

The Thames Tideway Tunnel is the UK water industry's largest ever infrastructure project. It is part of a three-stage project to cut river pollution and clean up the Thames:

- Stage 1 upgraded the main sewage treatment works in London – plants including Mogden, Crossness and Beckton can now treat more sewage every year.
- Stage 2 built the Lee Tunnel. Operational since 2015, the 6.9 km tunnel reduces sewage overflowing into the river Lee.
- Stage 3 is the Thames Tideway Tunnel. The tunnel is 25 km long, 66 m deep and 7 m in diameter. Due for completion in 2025, the tunnel will connect with 34 of the most polluting overflow points along the river – collecting sewage that currently overflows into the Thames and transfer it to Beckton for treatment. Engineers used six tunnel boring machine (TBMs) to dig the tunnel. Each tunnel boring machine weighed 1 350 tonnes and was 147 m long – the same length as 12.5 double decker buses. Tunnelling started in 2018 and went on for 24 hours a day for four years to 2022. The project also created three acres of public space along the Thames – including at Blackfriars (near the City) and Victoria Embankment (near Downing Street). All the new spaces aim to let people get closer to the river.

A career in water and wastewater engineering offers the following benefits:

- The work you do on a day-to-day basis is vital, making a real difference to people's lives and to the environment. Your work will impact hundreds of thousands of people, sourcing water and supplying it to homes, farmland and other businesses. This can be extremely rewarding, especially in developing countries where reliable sources of water may be scarce.
- The work has huge variety and offers routes to operational, design and construction management.
- The water resources industry is filled with opportunities to work across the globe due to its universal reach. Major world issues such as climate change and urbanisation are presenting greater challenges than ever – from providing irrigation water to farmers in the developing world to building resilience into water and wastewater networks, in some of the most advanced water systems ever constructed.
- With an increasing awareness of public health concerns, there are also new challenges to be met in the design and operation of sewerage and wastewater treatment systems to meet increasingly demanding standards necessary for the protection of rivers, estuaries and coastlines from pollution.
- A rapidly changing area, there's opportunities to work with exciting technological developments such as smart infrastructure management systems that monitor leakage and water quality; state-of-the-art drone technology and unmanned aerial vehicles to provide real-time flood monitoring, and digital twins that digitally represent assets in the environment and simulate 'what if' scenarios relating to flooding.

Harriet Cheaney CEng MICE
Construction Engineer/ NEC Supervisor, Jacobs

Inspired by industry experience

When I'm asked what inspired me to become a civil engineer, I tell them I signed up for a civil engineering degree because it interested me – the combination of logical thinking, maths, physics and team working. I had no set plan to pursue a career in civil engineering, I was happy to see where the degree took me.

It was only when I got on site during a summer internship working on Crossrail that I realised how fantastic civil engineering really is.

I love how many career opportunities there are in civil engineering. You can start in one role and easily traverse across to other roles depending on what interests you want to pursue in your career journey. Every day is different and you can be proud of what you work on.

Tideway: increasing public appreciation of engineering

One of the highlights of my career has been when friends and family recognise the projects I've been working on.

Tideway has been a fantastic example where the publicity has given the public a better understanding of what engineering and construction is all about, and how the scheme will help to

clean up the River Thames.

It makes me feel proud not only for working on a project which I fully believe in, but also for being an engineer.

A flexible career

Currently I'm on maternity leave. I plan to take a full year off and return to a part time role while my son is still young.

The new way of working since the pandemic has been transformational, especially for project and site-based roles. This has been invaluable for my return to work so I can continue my career alongside having a family.

Before going on maternity leave I was based on site where I worked in an assurance role for the client. I was responsible for checking the works were being carried out correctly by the contractor. I love being on site, it is fast-paced and you get to see everything.

> I love how many career opportunities there are in civil engineering. You can start in one role and easily traverse across to other roles depending on what interests you want to pursue in your career journey.
>
> Harriet Cheaney

Since COVID, there is now the possibility to work from home which is great as it allows time to focus fully on getting my work done, as well as a break from the early mornings!

One of the key changes I've seen in the profession is in diversity, this being me for example. A working mum who is able to work part time on a project-based role – fantastic.

Getting chartered: a great feeling of accomplishment

My greatest achievement has been getting chartered with ICE. I had prepared to sit my review just before the first COVID lockdown and was determined to get it done.

When ICE rolled out online reviews I was one of the first to sit my review successfully. It was a great feeling of accomplishment, especially while the rest of the world was still in lockdown.

Professional qualification has helped me the most when planning my return to work during maternity leave. All the hard work to sit my ICE review meant I had neatly summarised my key skills and achievements and boosted my confidence on what I have accomplished already.

Loch Ness: a civil engineering story (Galliford Try)

Building a water treatment works on a tranquil hillside by Loch Ness may sound straightforward, but it doesn't come without its challenges! The Loch Ness project has stretched the problem-solving abilities of its engineers, innovated world firsts, crossed international teams and garnered several industry awards for the solutions forged during its development.

The project

Delivering on behalf of client Scottish Water, the ESD joint venture comprising Galliford Try, MWH Treatment and Binnies, developed a new solution for water treatment for the communities of Fort Augustus and Glenmoriston by Loch Ness, at the heart of the Great Glen.

The original scope involved the construction of a new intake system from Loch Ness up to a new treatment works with capacity for 1 million litres of water per day. This would feed two service reservoirs – one at Invermoriston, and the other 8 km away on a hillside above Fort Augustus (connected by pipes running along the famous Great Glen Way).

The team used a new digital delivery model (Civil 3D) which allowed designers to visualise the land, react in real time to conditions on site, as well as to allow continuous improvement to the design as the project progressed. The planning incorporated a carbon reduction ethos and a mission to reuse as much waste material as possible – as part of our ESD JV, 98% of site waste is currently reused or recycled.

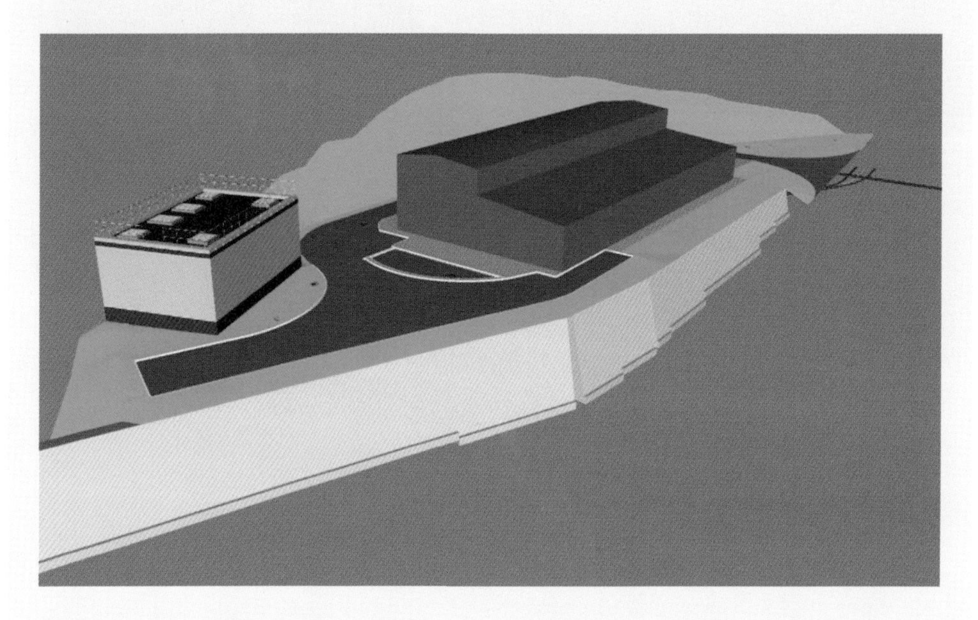

Challenge 1: The installation of 75 metres of pipes in Loch Ness with minimal environmental disruption

In order to meet the brief, the project team had come up with an alternative solution for keeping the pipes in place due to the fact it wasn't possible to assemble them or pour concrete, beneath the loch's waves.

Solution: The team used polypropylene mesh bags filled with rocks to form the main structure for water intake – eliminating the need for concrete. In addition, the pipes were welded on land – rather than assembling at site – and then floated into place using a ballast with concrete collars positioning the pipes.

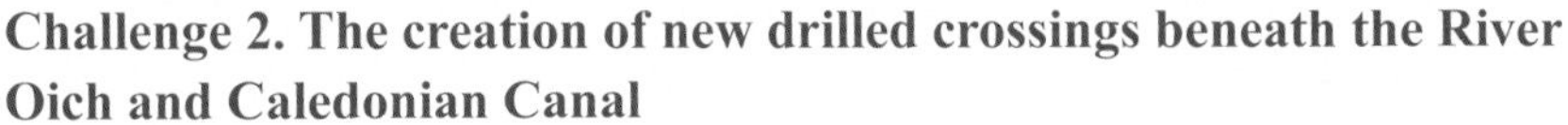

Challenge 2. The creation of new drilled crossings beneath the River Oich and Caledonian Canal

There was a requirement to use horizontal directional drilling and pipe borings for 350 metres of distribution pipes that would run under the River Oich and the Caledonian Canal, supplying the local population with clean water from the plant.

However, drilling underwater requires the use of a special fluid (or mud) that removes cuttings from the hole, cleans the drill bit and powers equipment like motors, and soon after starting to drill beneath the canal, the team discovered that the drills were losing fluid with no obvious explanation. The next step was to investigate and work out a solution.

Solution: It turned out that the builders during the 19th century had used large broken rocks to fill the voids under the canal during construction, meaning the drill was not just going through solid rock, but also hitting pockets of water. Through collaborating with specialist Finnish company GeoNex, the project team found a solution that had never been tried before in Scotland and had never been used for such a long distance.

The approach was to add a steel casing around the drill, essentially creating a closed tube for pulling the pipes through. The casing offered stability when drilling through unpredictable layers of rock, prevented fluid from leaking out and had the bonus of offering strong protection for the pipes themselves.

Could you meet the challenges?

This is just one small example of scenarios our teams work through on a daily basis to deliver efficiently – if you find this combination of physics, problem solving, environmental management and technological challenge interesting, a career in our industry may just be for you!

Matthew Naylor GMICE
Site Agent, JN Bentley

How would you describe your current role? What do you enjoy about it?

I'm currently a Site Agent working on a flood defence scheme at Clementhorpe in York. It's a very busy site with a lot of people involved (such as residents, York City Council, the Environment Agency) and with multiple work areas. I enjoy the challenges that come with being a site agent – no two days are the same and I often find myself problem solving design challenges, such as working around existing structures. Managing a number of people and interacting with site visitors means I'm always hands-on and working in a dynamic environment, interacting with a range of people, from the client to JN Bentley commercial managers and operations directors, to ground labourers.

What stands out as the most interesting project you've worked on?

I was based at Great Yarmouth – one of our largest projects for the Environment Agency – where we carried out major works to renovate and replace flood defences. I worked first as a site engineer and then as a site agent on works package 18. The project won the Sustainability Initiative Award at the British Construction Industry Awards, which was to do with the sheet piling – we created a new defence line using steel sheets, which saved a lot of money and carbon, the design of which I had involvement with for a number of months. It was an innovative and bespoke solution to ongoing issues in the quayside, and the marine engineering aspect was really interesting as it's something we don't typically do.

Career development, education, first role, and career highlights

I did a civil engineering degree at Derby University, then a placement year in Leeds working as an assistant site manager on hotel construction. I wanted to move into infrastructure development, so joined JN Bentley as a site engineer in 2017. In this role I worked on a range of projects which helped me to build my technical and practical knowledge – I had the opportunity to work on flood defence projects for the Environment Agency and mine water treatment schemes for the Coal Authority. I then moved into a project designer role in JN Bentley, working on the design for a flood alleviation scheme at South Ferriby for two years, moving between site and design, which helped me gain lots of experience and confidence.

I'm currently working towards IEng at the minute and looking to do my professional review this year. I've also been put through a civil engineering law and contract management course over the past 6 months at Leeds University, so I have two ICE exams coming up soon.

As I've built my career, I've found that other site agents and engineers have been really good at coaching me – as I've progressed I've built valuable relationships and a support network of people I can turn to for advice.

What changes in the profession you have seen and where do you think the profession is heading?

Since I've been at JN Bentley, the health, safety and environmental awareness culture has developed massively – something that the industry as a whole is developing, but our

'beyond zero' target means we are aiming beyond the curve and striving to be at best practice level within everything we do.

I've also been part of a huge carbon push over recent years – on my first project, I didn't really know what carbon meant, and now it's a forefront issue and a consideration that informs all of our decisions. We consider where we source materials from and what type of materials, something we did exceptionally well at Great Yarmouth. We're also better at mitigating our impact (such as replanting trees) which is where our OSSs and OESs come in which have been great developments. Our integration with JBA and Mott MacDonald is a huge benefit as the design and build element means I have the resources available to get an environmental advisor or agriculturalist out to site, a great positive for us and I can see collaboration becoming more important to the industry as opposed to traditional ways of working.

What's next...

There is a broad range of opportunities for graduates entering the water sector in the fields of water distribution, treatment and management, wastewater treatment, collection and management, hydrology, flood management, monitoring of water bodies, integrated water resources management and other areas.

Graduate engineers often find employment in consultancies, with utility contractors, environmental agencies, environmental and water engineering consulting firms, construction companies carrying out water engineering works, and in government departments. Many organisations offer opportunities to gain experience across a range of difference types of project, for example working across wastewater projects, clean water projects, and flood defence and alleviation projects.

02

Disciplines of Civil Engineering

Coastal and maritime engineering

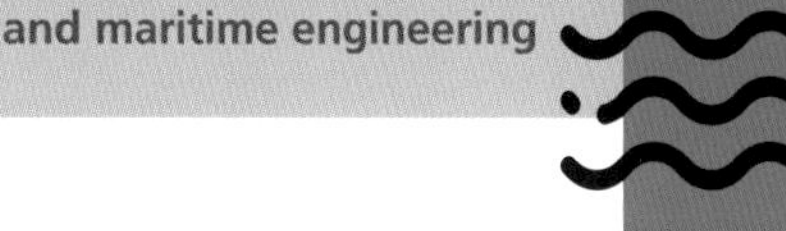

Coastal and maritime engineering

Coastal erosion, flooding and sea level rise, as well as infrastructure such as ports and harbours all require skilled input from civil engineers. Practitioners in coastal and maritime engineering help communities and businesses to understand, maintain and design sustainable coastal environments.

Coastal and maritime engineering is where civil engineering meets the sea. The range of projects in this area includes: marine infrastructure along the coast and offshore; beaches and the movement of sand; protecting developments along the coast from the effects of wave action, changes in sea level and currents; and determining safe setback lines for new developments. This sector also includes all aspects of civil engineering relating to ports – from infrastructure design, ship navigation channels and dredging, to port operations and layouts.

Ports and maritime

Ports and harbours play a vital role in the transportation of people and goods across the world. Many of the things we rely on in our daily lives – the food we eat, the clothes we wear and the energy we use – travel through ports. Civil engineers working in ports and maritime are involved in projects that span the entire project lifecycle of ports and harbours, from planning, design and construction, to the operation and maintenance of maritime structures and infrastructure. Work in this sector will often involve design, development, refurbishment and construction of ports, port infrastructure, bulk/container terminals and coastal structures, such as quay walls, piers and jetties, and breakwaters. Some recent trends include: the development of smart ports, use of digital twin technologies for predictive maintenance and predictive berthing, cargo flow optimisation, and planning port infrastructure for the introduction of autonomous sailing vessels.

Liverpool2

Liverpool2 is one of the UK's largest private sector infrastructure projects, developed in response to changing patterns and shipping trends towards the use of 'mega' ocean-going container ships.

It required the construction of a deep water quay 854 m long and 30 m deep with an adjacent container terminal of around 16 hectares reclaimed from the mouth of the River Mersey. Two new berths now allow access for the latest generation of post-Panamax vessels which can carry loads of up to 135,000 containers.

The new terminal location is close to the Manchester Ship Canal. There are ten motorways within 10 miles and ten rail linked terminals within the port estate.

Project achievements and benefits

Liverpool2 can handle the largest cargo vessels in the world. Global shippers have a viable alternative, able to transport cargo more efficiently to their end destination. The result is a sustainable logistics option, enhancing regional economic growth, reducing transportation cost, vehicle congestion and carbon emissions.

The river deep berth terminal concept was originally conceived over a decade ago. Procured in 2012, the first piles were pitched in September 2013. Land reclamation started in February 2015, ship-to-shore cranes were off-loaded in November 2015 and the grand opening ceremony took place on 4 November 2016.

The existing dock entrance dock (built in 1905) is not large enough for mega container vessels, so the new quay was located on the River Mersey enabling access to the largest vessels. One of the tallest quay walls in Europe was required – to accommodate the draft of the vessels and 10 m tidal range – standing at 30 m high from the top to the bottom of the berth pocket. The 854 m-long quay wall was formed by installing 13,500 tonnes of steel and retains land that was created from 5.5 m^3 of sand dredged from the Mersey.

Coastal engineering

Engineers play an important role in developing sustainable solutions for coastal, estuary and river systems, and managing the increasing risks brought by climate change. Coastal engineering and management involves a range of disciplines related to the interaction of the sea with the land. In changing the way that we

manage the shoreline, engineers must take heed of the natural processes at work to maximise the desired outcome at the project site and further afield. Coastal engineering and management not only involves construction in coastal areas and the project stages leading up to that, but increasingly it requires understanding the implications of letting nature take its course.

Traditionally, coastal engineering has involved building sea walls or other measures to prevent sea floods, preventing erosion of cliffs, and improving the stability of coastal slopes. Work in this area is often divided into hard and soft engineering approaches – hard engineering methods include the design and construction of seawalls, groynes, offshore breakwaters, floodgates, rock armour and revetments, for example. Soft engineering techniques involve working with nature to manage the coastline. This includes cliff stabilisation, dune regeneration, beach drainage systems and the creation of buffer zones. Roles in this area require a detailed understanding of the complex and highly variable interactions of the atmosphere, ocean and adjacent lands, and over wide time scales. So although coastal engineers are usually qualified in civil engineering, they must also understand the principles of oceanography, geology and other aspects of the marine environment. Coastal engineering combines a broad scientific knowledge with the practical skills needed to plan and implement the infrastructure that meets the needs and expectations of society.

Professionals in this area often highlight the following benefits:

- The work of coastal engineers is vital in safeguarding coastal communities that are most at risk from flooding. The benefits of your work are long lasting, and the designs will be used and operated for many years into the future.
- The projects you work on make a real difference to the environment. Development at the coast can have huge impacts on coastal habitats, so effective planning and management with the environment in mind is crucial.
- The role and projects themselves vary widely so that no day is the same. New challenges arise with each project, providing a stimulating work environment.
- Frequent visits to the coast are also a bonus!

Great Yarmouth tidal defences (JN Bentley)

Great Yarmouth has a history of flooding. As recently as 2013, 9 000 people were urged to evacuate their homes when the highest ever tide was recorded on the town's river.

With the existing defences along the River Yare and River Bure requiring improvements, we were asked to design and construct a solution that would strengthen flood protection to over 4 500 homes and businesses.

Rather than use the same conventional approach across the whole 4.5 km stretch of repairs, we carried out testing that meant we could implement innovative solutions specifically suited to each section.

This resulted in us engineering a combination of plating, wall raising

and cathodic protection along the river, as well as building new sections of walling set back from the waterfront. It also meant that where repairs weren't required, we didn't need to carry out any unnecessary work at all!

By targeting the repairs, we scaled back the original extent of work needed and made a major carbon reduction, far in excess of our 40% target.

Working closely with specialists, limpet dams created a safe working area below the level of river water, enabling the plating to be installed safely.

We teamed up with the local supply chain to deliver the project efficiently. The manufacturer of the new precast concrete walls was based just 150 m from site; steel fabricators and workboat providers were based in Great Yarmouth; and our marine experts were based down the road in Norwich.

Now complete, the £30m+ project is one of the largest we have ever delivered with the Environment Agency. The solution has not only bolstered the flood defences in Great Yarmouth, it has done so in an affordable and low-carbon way. The project has since been recognised with a series of awards, including winning Environment and Sustainability Initiative of the Year at the British Construction Industry Awards.

Working closely with specialists, limpet dams created a safe working area below the level of river water, enabling the plating to be installed safely.

Typical roles and activities in coastal and maritime engineering include:

- **Port planner** – Development of studies across project phases to ensure commercial and technical feasibility. Develop port planning tools, model port development capacities, develop masterplanning optioneering and develop proposals and commercial offers.
- **Maritime civil engineer (Graduate / Senior / Principal)** – Responsible for completing detailed design calculations, concept designs and feasibility studies. Opportunities for site supervision work and inspections dependent on projects.
- **Coastal modelling** – Computational and physical modelling of wave hydrodynamics, structures and interactions to support designs.
- **Port and coastal structural engineer** – Design and review of the structural infrastructure of port and coastal developments, including reclamation, jetties, breakwaters, promenades, wharves and docks.
- **Project manager (coastal and maritime)** – Responsible for managing all members of the team to deliver a project. Liaises with the client, organising meetings and running the project on a day-to-day basis.
- **Marine consenting** – Interfaces with key consenting stakeholders such as port authorities, the Environment Agency and the Marine Management Organisation to obtain permisions and consents for developments.

Roles in coastal and maritime engineering often involve:

- Hands-on involvement on projects, for example, with inspection, design, feasibility, tendering, construction, management, and maintenance
- Developing a thorough understanding of relevant standards, codes of practice and current design methods
- Computational modelling of coastal environments
- Environmental assessment and conservation
- Using technical engineering skills to assist in the design and analysis of complex maritime structures, and in the preparation of technical reports, drawings, calculations, risk assessments and method statements
- Contributing to work winning and preparation of bid documents
- Helping to prepare contract documents, bills of quantities and specifications
- Maintaining relationships with a variety of clients and suppliers.

Practitioners often cite the following benefits of careers in coastal and maritime engineering:

- A chance to create, improve and protect the environment in which we live. Maritime and coastal engineering are critical to the global supply chain, energy supply, the construction industry and for safeguarding coastal living.
- Projects with a variety of clients, locations and workplaces, all with real-world impact.
- Structures in the coastal environment are fascinating as they need to balance lots of factors – you'll need to consider ground conditions, wave and tidal conditions and lots of further environmental conditions such as wind speeds, sediment dynamics, etc.

- No two projects are the same, due to existing constraints and the individual nature of the structures designed.
- Opportunities to work internationally.
- Practitioners in maritime and coastal engineering interface with many other professionals in a collaborative manner i.e. with economic advisory on market studies, highways on port road networks, building structures on warehousing, as well as diving technicians, materials scientists, geomatics specialists, and hydrographic surveyors.

CAREER PROFILE

Jamie Arnott EngTech MICE
**Graduate Engineer,
Wallace Stone LLP**

A good sense of job satisfaction

As a Graduate Engineer with consulting civil engineers Wallace Stone, I am responsible for undertaking design duties and producing technical drawings for projects within the field of marine and floating structures.

I enjoy working within this field as each project within the marine environment presents its own unique challenge – no two jobs are the same. There is also a good sense of job satisfaction when a project is successfully completed and the client is happy.

Efficient planning and good communication

The most interesting project I have worked on is the Tarbert Ferry Terminal Redevelopment project for Caledonian Maritime Assets Ltd.

This complex project involved the demolition, replacement and extension of the existing pier structure in Tarbert on the Isle of Harris in Scotland. The scheme also included the formation of an extended ferry marshalling area, dredging of the seabed and future construction of a new ferry terminal building.

A key challenge was to complete the works around the uninterrupted operation of the ferry service. A particular focus had to be placed on the design of the temporary works and phasing of all demolition and construction work to minimise disruption.

Thanks to efficient planning and good communication between all project participants, only one ferry sailing was affected throughout the entire two-year construction programme.

A passion for drawing and design

My passion for drawing and design led me to pursue a career in civil engineering. This passion stems back to when I was younger and received a drawing board as a gift. At high school I was introduced to computer aided design and I realised I had the ability to turn my hobby into a career.

Civil engineering offers a wide range of opportunities, career paths and a variety of sectors, with the chance to work on projects which positively contribute to the built environment.

The varied nature of the profession means that it requires people with different skillsets and experiences to work together. This variety and demand for varying skillsets means that civil engineering can be accessible for all.

Degree apprenticeship

I have recently completed a graduate level apprenticeship in civil engineering at the University of the West of Scotland.

This course has given me a BEng (Hons) in civil engineering within four years, which is the same length of time it would take to complete on a full-time basis. I attended the university campus one day per week and spent my remaining four days per week working in the office.

The apprenticeship has allowed me to progress professionally by expanding my knowledge and accelerating the time it took me to gain professional accreditation. The thing I enjoyed most about the course was that I was able to learn new skills which were relevant to my role and field within civil engineering, allowing me to take on more responsibility within the workplace.

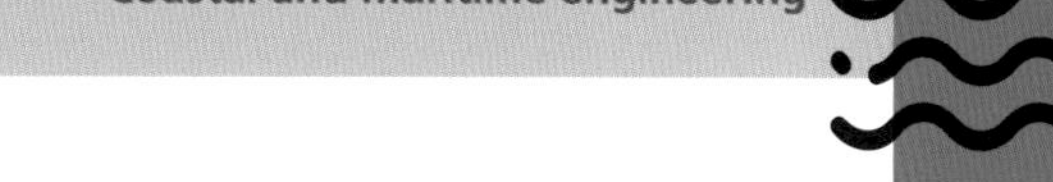

A set standard

I was motivated to become professionally qualified as I was keen to progress within my role and have opportunities to advance my career in the future. I also wanted to measure my ability and receive recognition for what I had achieved.

Becoming a professionally qualified member of ICE has helped me in my career as it recognised that I had reached a set standard of professional work. This gave me a renewed confidence in my ability and has encouraged me to continue developing and gaining knowledge.

> Civil engineering offers a wide range of opportunities, career paths and a variety of sectors, with the chance to work on projects which positively contribute to the built environment.
>
> Jamie Arnott

Key skills for coastal and maritime engineering include:

- You'll need a good understanding of basic engineering principles. Many of the engineering structures designed are large ('heavy civils') and go beyond what's covered in standard engineering design codes used in other sectors.
- Problem solving is a large part of the work and it's important to have a drive to solve difficult problems with a methodological approach – perseverance is key to working through to find a solution.
- As with all disciplines of civil engineering soft skills are becoming increasingly important – this includes being able to form and maintain good working relationships and inspire and motivate others, engaging productively with stakeholders, resolving disputes, and negotiating effectively. Other key skills include time management, creativity, leadership, attention to detail, budget management and critical thinking.
- Good written communication and attention to detail are also important, including accurate technical analysis and good report writing.

Next steps...

A large number of coastal engineers has a bachelor of science in civil engineering, although other environmental or scientific backgrounds feature heavily and are often complemented with relevant postgraduate studies, for example an MSc in coastal engineering. Graduates can then join coastal engineering firms to develop specialist skills and understand the sector through on the job training.

Graduates generally have the opportunity to work on a wide variety of projects from the start; and due to the nature of maritime in particular, there can also be opportunities to be involved with a range of different types of work as well. For example, graduates may be working on feasibility studies and writing reports summarising solutions, or they may work on detailed design calculations for a structure due to be built in the near future.

There are often opportunities to see construction on site and take part in supervision activities. Graduates also get involved with inspections of existing structures on site, followed up by the development of inspection reports and recommendations of actions to take.

02
Disciplines of Civil Engineering

Transport

Transport

Civil engineering careers in transportation span the planning, design, construction, management and maintenance of facilities for all types of transport. It's a broad and ever-changing sector, which develops as the needs of society and businesses change.

Civil engineers are involved in all modes of transportation and so careers in this area span a broad and diverse range of roles. Career pathways fall broadly into two categories: roles working within a planning and development context, and roles in an engineering design and infrastructure context.

Transport planning

Transport planning focuses on when, where, how and why people and goods move around in the built environment. Transport planners work at the earlier stages of the design and engineering process, defining and understanding objectives and solving the problems and constraints such as capacity and congestion, decarbonisation and modal shift, journey times and reliability, social mobility, accessibility, health and wellbeing, levelling up, diversity and inclusion, environmental impacts and sustainable growth. The focus is on the needs of users to define strategies and business cases, and helping clients understand elements such as the policy and regulatory context, and funding requirements to lead them through the design process.

Transport planning as defined by the Transport Planning Society is divided into six core areas:

1. **Policies and regulations** including understanding governance, procedures, planning requirements and financing in a national, regional and local context.

2. **Tools and techniques** including the collection, processing, analysis, managing, reporting and visualisation of data, especially to support detailed modelling and forecasting for transport schemes using analytical software.

3. **Planning and design** including the preparation, development and delivery of transport plans, designing for transport integration, accessibility, security, sustainability and other scheme specific objectives.

4. **Operations** including understanding, planning and designing for changing travel behaviours, particularly relevant in a post-pandemic world, and commercial operational management.

5. **Management** including understanding project management principles such as procurement, project proposals, time management, cost, and scope.

6. **Communication, commitment and ethics** including the ability to visualise, report, present and discuss the outcomes of work with internal and external teams.

Transport planning aims to maximise connectivity for people and businesses, while minimising the need to travel, to achieve movement of both people and goods as sustainably as possible. In densely populated areas, rail and bus-based public transport systems can be used effectively to limit some of the congestion that large numbers of vehicles can cause. Trains, planes, and ships can be used for transport across increasingly long distances in an orderly and safe fashion. Recent developments include planning infrastructure for electric and hydrogen powered cars, and for autonomous vehicles.

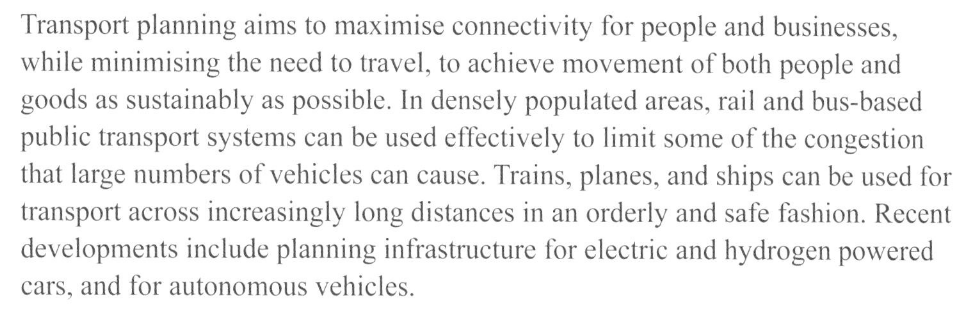

Transport engineering

Transport engineering is the application of technology and engineering principles to develop innovative design solutions for transport infrastructure, including tunnels and bridges for railways and highways, and other infrastructure such as airports and seaports. This field of engineering involves working with transport clients to design infrastructure that is safe, economical and sustainable, as well as meeting all the client requirements. Transport infrastructure is important to increase the interconnectivity of cities and communities.

Transportation engineering encompasses numerous specialisms focusing on infrastructure development to increase spatial connectivity. For example, specialist design of highways is fundamental for well-designed, well-built and well-maintained roads, to ensure effective and efficient infrastructure that is safe for users.

In development of rail, highways and aviation schemes, physical obstacles will inevitably present themselves. When these cannot be moved, engineers may decide to use a tunnel or a bridge. These elements of transport engineering require their own specialisms.

Typical roles across this sector include:

- **Transport planners** – when a public or private organisation would like to develop a scheme, a transport planner can support in understanding what the baseline conditions are and what needs to change in order to meet the scheme's objectives. This includes undertaking transport assessments, understanding user needs, land use, access requirements, safety considerations and value for money.
- **Transport modellers** – often when it comes to the road and rail network, bespoke software packages are required to model and forecast demand to make sure that the transport network has sufficient capacity to operate effectively. Strategic modellers work over a wide geography, ensuring the strategic network connecting cities and towns is not congested, and process and analyse large amounts of data to understand constraints.
- **Highway engineers** – when designing, a highway engineer must be aware of the area they are designing in. For example, a road being designed in a residential area may need to have measures to mitigate the visual and acoustic impact on the residents, whereas a motorway that isn't in proximity to residences may not have to implement these same measures. In addition, a highway engineer must ensure that their designs are safe for the specified user and fit with local road networks.
- **Rail engineers** – it is of the utmost importance that a rail engineer understands the societal costs and benefits of a project. A new railway may have properties and habitats in its path;

therefore, rail engineers must find creative solutions to working around these obstacles. Rail engineers must also liaise with stakeholders as the new rail will impact them directly.

- **Tunnel engineers** – a tunnel engineer must ensure that thorough surveys are undertaken, and that the information received is accurate as once tunnelling works begin, delays could be detrimental to a project. It's important therefore that information regarding obstructions and geotechnical make up is collected and considered. A tunnel's main purpose is to limit the impact on existing infrastructure and it's the engineer's responsibility to ensure that the construction and operation of the tunnel will not cause undue disruption.
- **Airport engineering** – a civil engineer in the airports sector will specialise in the design, tendering, contracting and supervision services in connection with runways, taxiways, aprons and landside access facilities. The civil engineer must consider issues such as drainage, airside layout, landside access design and passenger flow characteristics.
- **Port engineering** – ports remain a vital part of the global infrastructure systems and have very special engineering considerations due to tidal movements. As with all maritime projects special consideration is needed to consider the salty marine environment which affects the life of materials. Specialist areas of port engineering include metocean and hydrodynamic investigations, port masterplanning, feasibility assessments, through to detailed design, design and build, construction documentation and site supervision.

Within the transport sector there are also opportunities in:

Structural engineering – design of substructures and superstructures in transport systems.

Urban design – design of street furniture, and urban planning and design more generally, including green infrastructure to create places for communities to enjoy.

Project management – management of engineers on a project to ensure complex systems are delivered.

Roles in this sector often involve:

- finding new solutions to transport problems
- using computer-aided design software
- ensuring the safety and legal compliance of engineering work
- planning and supervising new projects
- ensuring that projects are delivered on time and within budget
- offering engineering guidance to other team members
- dealing with key stakeholders including clients, transport specialists and members of the public
- preparing tender and contract documents.

CAREER PROFILE

Calum Farquharson
IEng MICE
Senior Highways Engineer, WSP

A fascinating variety of projects

As a Senior Highways Engineer, I have been involved in a fascinating variety of projects, including designing new and maintaining existing infrastructure such as carriageway resurfacing, footway improvements, drainage improvements, safety barriers, pedestrian crossing upgrades, sea wall projects and rock slope stabilisation projects.

I typically work on multi-disciplinary projects with other specialists so I can find the best solution to my problem and ensure I make as many improvements as I can to the surrounding environment, rather than just focusing on a singular issue.

I have had some really interesting experiences in my career so far. I had the privilege of undertaking a student placement on the new Queensferry Crossing in Scotland. I worked on a complex drainage project in Australia, and oversaw the excavation and installation of an outdoor swimming pool which I found quite fascinating.

In 2018, I delivered a complex carriageway resurfacing project on the west coast of Scotland. This involved unique design proposals, substantial health and safety measures and an extensive public consultation process. This project is one of my highlights because it was in a dense urban location with numerous hazards and the potential for major disruption to the local community. After successfully delivering the project, I received a personal thanks in the local newspapers, recognition from local councillors and received an award from the company director that I worked for at the time.

My largest project to date was a coastal defence scheme, where I was lead designer and site supervisor. The £1.4m project was located on the west coast of Scotland and completed in 2020. The works involved excavating and replacing the existing 145m long sea wall, destroyed during the storms of winter 2018–19. As part of the works, I also upgraded some of the surrounding highway infrastructure including the footway, safety barrier and drainage.

Incorporated Engineer: the biggest highlight of my career

Without a shadow of a doubt, the biggest highlight of my career to date was passing my review to become an Incorporated Engineer with ICE (IEng MICE). This is number one for me because it was the culmination of years of experience I had built up. To discuss my experience with two esteemed professionals, and for them to consider me competent enough to be admitted into the institution, was a real privilege.

It can be a daunting prospect at first, but the process to become professionally qualified is worthwhile. ICE has an excellent membership development team that is always on hand to guide you. Becoming professionally qualified with such a prestigious institution was always a goal of mine, and passing my IEng review presented me with opportunities that I may have otherwise been overlooked for. It has given prospective employers and clients that don't necessarily know me, confidence in my ability.

'Life is like grass'

An old boss once said to me 'life is like grass, if you're not growing, you're dying'. That's stuck with me throughout my career so far, I think it's a great analogy for any aspiring engineer.

I was quite nervous when I started working in the industry. Coming out of university, I didn't know that much in terms of

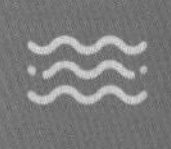

designing solutions to real-world problems. I was worried that there was an overwhelming amount to learn and that I would struggle to prove my worth to an employer.

What I quickly found was that in this industry, we are all constantly learning, no matter how many years of experience you have. In fact, two of the key attributes required for gaining a professional qualification with ICE are identifying the limits of your knowledge and carrying out continuous professional development. Being a member of ICE also gives you access to some fantastic learning resources.

We are all learning – it's nothing to be afraid of.

> Becoming professionally qualified with such a prestigious institution presented me with opportunities that I may have otherwise been overlooked for. It has given prospective employers and clients that don't necessarily know me, confidence in my ability.
>
> Calum Farquharson

Key skills in transport roles include:

Transport planning

- Excellent problem-solving skills and, equally important, an ability to understand the human element of transport planning.
- Being organised and efficient in order to meet deadlines and complete tasks effectively.
- Good analytical skills and an attention to detail, particularly for transport planners who carry out modelling work to ensure models are correctly projecting future scenarios and proposals.
- Clear and precise communication skills are at the heart of any engineer but particularly so for a transport planner.

Transport engineering

- Good problem-solving skills and creativity to be able to develop innovative solutions to engineering problems.
- Effective communication is also required to convey ideas to colleagues and clients through presentations, as well as to document the design process clearly in design and calculation reports.
- Time management and organisational skills are also important to ensure client deadlines are met and the design programme is carried out with minimal delays.

The Elizabeth Line

The Elizabeth line is a new railway through the UK capital. With London's population set to reach 10 million by 2030, it's designed to improve a public transport system already struggling to cope.

It connects the city of Reading in Berkshire and Heathrow airport in west London, to Abbey Wood in south London and Shenfield in Essex. Linking 40 stations – ten of them new – the line provides the biggest increase in central London's train capacity ever delivered by a single engineering project. It is expected to see 200 million passenger journeys every year.

The project dug out 42 km of new tunnels under London using eight 1 000 tonne boring machines. This part of the work took 3 years and used over 200 000 concrete segments to create the new 7 m diameter tunnels.

The route for the tunnels was a major challenge for project engineers. They had to weave their way through existing underground railways, cable ducts and gas pipes. In some areas, they were less than half a metre from working tube lines.

Engineers also had to be very careful not to disturb the foundations of historical buildings along the planned route.

The project created 14 km of new station concourses, underground caverns and shafts using sprayed concrete lining. Some of these are the largest excavated spaces ever built.

The new railway is designed to be future proof. The stations are large with platforms 250 m long, and there is capacity for more space so that trains can be extended when it's needed.

Georgia Thompson
CEng MICE
Design Manager, BAM Nuttall

A passionate advocate for inclusion

I am a passionate inclusion advocate, so you can always find me getting involved in initiatives to make people feel more included – not only at work, but also in joining the profession. The industry is becoming much more conscious and socially aware and I love that.

One of the highlights of my career was meeting Her Royal Highness Princess Anne as a Patron of WISE (Women in Science and Engineering) in January 2021.

I was also proud to be selected as one of the top 50 women in engineering for my work in innovation.

A rewarding and varied career

My role as a Design Manager involves interacting with a lot of different people, solving problems and driving efficiencies. I sit between the design and construction teams.

I normally have three or more meetings a day establishing how our team is performing to programme. Sometimes there are gaps in information or problems that need solving. A big part of what I do is translating the information from one team, consents for example, to another, such as the client, and making sure the right people are in the 'room' to make the correct decision. In areas that need improvement, I create and

implement strategies to improve the process and support my team in delivering engineering designs and our construction teams out on site. I speak to a lot of different people to establish what solutions will make the biggest difference to our teams, and support them to deliver great infrastructure.

Plus, I always find any opportunity to get suited up in orange and go out to see some stuff get built.

A career in civil engineering is rewarding and varied. I love that each day is different and I can see my work being built in front of me. From a drawing to a live structure, and that is such a beautiful feeling.

A badge of honour

Being a member of ICE exposes you to different ideas and concepts outside your day-to-day professional life. I value learning how other people do things, better and more efficiently. It is a great vessel to be a part of and is making things better for future engineers.

> I am a passionate inclusion advocate, so you can always find me getting involved in initiatives to make people feel more included – not only at work, but also in joining the profession.
>
> Georgia Thompson

I have really enjoyed getting involved with the management side of the institution through my role a graduate representative. I get opportunities to work on projects that are changing the industry for the better and provide my insight. It is a great opportunity to meet people within the institution and discuss key themes that you are interested in. Learning about all the great things we do has been enlightening.

I successfully achieved my professional goal of becoming a Chartered Engineer (CEng MICE) with the ICE in July 2022. Chartership is like a badge of honour to me and official recognition of what I know. The journey towards achieving chartership was a collection of my experiences and how I have improved the work we do. It is something I am very proud of, and it has given me a level of confidence that I never had before.

Mimi Nwosu GMICE
Civil Engineer, Heathrow Airport

A rewarding role

I am a civil engineer at Heathrow Airport within the Engineering and Baggage function in the Technical Services team.

The Technical Services team at Heathrow Airport are custodians of all the civil engineering assets. This includes 5 terminal buildings, 4 million m2 of pavement, 44 road and rail tunnels, 125k roads, 200+ non-terminal buildings, and several specialist asset systems such as the control tower. I am a civil engineer within this team, predominantly focusing on the buildings and fabric assets. My duties include providing specialist technical knowledge to project managers, terminal teams, and operational teams.

My duties are to deliver continuous improvement from Heathrow assets over their whole life by implementing and assuring asset system strategies, standards, and plans. I work on several engineering systems across the airport, developing working relationships with operational, maintenance, development, supply chain, finance and safety teams. I am responsible for providing technical knowledge and support to operational teams within Engineering and Baggage function and the wider organisation to ensure robust asset management strategies are in place.

An iconic project and TV appearances

My most interesting project has been HS2. It is iconic and the largest infrastructure project in Europe.

I loved working on a joint venture, tunnelling is really fascinating and concrete is my favourite construction material. I worked as an assistant materials engineer learning to specialise in concrete quality and production.

My career has been extremely varied. I like how diverse the industry is and how easy it is to change sectors – I have worked in highways engineering for a consultant, materials engineering for a contractor and now in aviation for a client. Civil engineers will always be in high demand, no matter where in the world you decide to go. You can truly make your career your own and work on a variety of projects. I have loved using my knowledge and expertise in all my different roles.

I am proud of being named as one of the top 50 women in engineering by the Women's Engineering Society and being named in the Guardian newspaper. I have even had the chance to share my love for engineering with the world on TV programmes about engineering and construction.

Valuable resources and support from ICE

I am a Graduate member of ICE (GMICE), working towards becoming professionally qualified. One of the key motivations is to say I have reached one of the highest qualifications in my career and worked hard to achieve it. My managers and mentors also motivate me on the way, they are so skilled at what they do and technically excellent and I aspire to be the same.

Achieving professional qualification will mean I can climb the career ladder in my company and prove I am a competent engineer who completes work with integrity. I will be more confident in my engineering judgement to make big decisions on projects.

I value the resources and support I get from my ICE membership! If I want to keep my industry knowledge up to date, network with industry professionals and have any questions I have answered, I know I can get assistance from ICE. Having membership can help shape your career.

Professionals working in this sector often highlight the following benefits:

- It's a very diverse field which has every specialty you can think of from aviation, rail, highways, walking, cycling and wider development planning to name a few. The variety of projects can go from early-stage strategic cases for new transport projects through to modelling proposed schemes and producing detailed technical specifications.
- There is also lots of variety within different types of design projects, including feasibility studies, detailed design and design checks, or regeneration and repair works. The types of projects you could get involved in include the design of bridges, tunnels, ports, civil structures and much more!

- Transport planning is multi-faceted and the impacts of a scheme go beyond simple journey times – it is related to health, employment, life expectancy, communities and arguably more important than ever: climate change.
- Each project in transport engineering is often different, with its own unique challenges, making the work exciting.
- It is a great sector to work in to influence walking, cycling and public transport, and ultimately help lower carbon emissions. It's also becoming increasingly important to identify opportunities for reducing embodied carbon emissions through design for example.
- Transport infrastructure is needed globally so there are often lots of international project opportunities which are exciting.

CAREER PROFILE

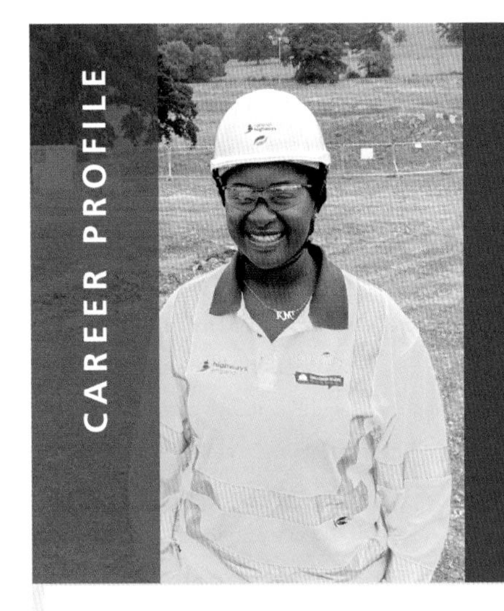

Moyin Pelumi, GMICE
Graduate Engineer, Galliford Try

Three career highlights

1. Seeing the first signpost I set out being installed.
2. Being assigned my first package to look after.
3. Learning how to make decisions on my own.

How did you end up in your role?

I secured a position as a Graduate Site Engineer with Galliford Try in the highways sector on completing my BEng in civil and structural engineering from Newcastle University. Since starting my job, I have been with Galliford Try for two years. As I have completed my graduate programme, I will be moving to my new role as an Assistant Planner to get a greater feel for the different aspects within the construction industry.

What does a typical day look like?

In my line of work, there's no typical day as every day is different but my main responsibilities entail:

- Setting out points, co-ordinates, and levels for sub-contractors so they can accurately build highways and structures.
- Use of GPS to check the work is being constructed according to design.

- Collecting daily diaries to keep track of subcontractor activity.
- Completing check sheets to make sure work meets the specification and updating the tracker to ensure we're on programme.
- Utilising the programme to plan ahead.
- Hosting the start of shift briefing and sending it out to the project team to make sure everyone is aware of the programme for the following day.

I would recommend a career in civil engineering because...

It is a fulfilling career especially when you see the final product of a project.

What's the biggest/most complex thing you've made out of Lego? How long did it take you?

During my graduate programme, we were given a task to build the tallest, most stable building out of Lego. The most challenging part of the project was making the base stable enough so that it could withstand the weight of the extra block we would periodically place on top of the structure to

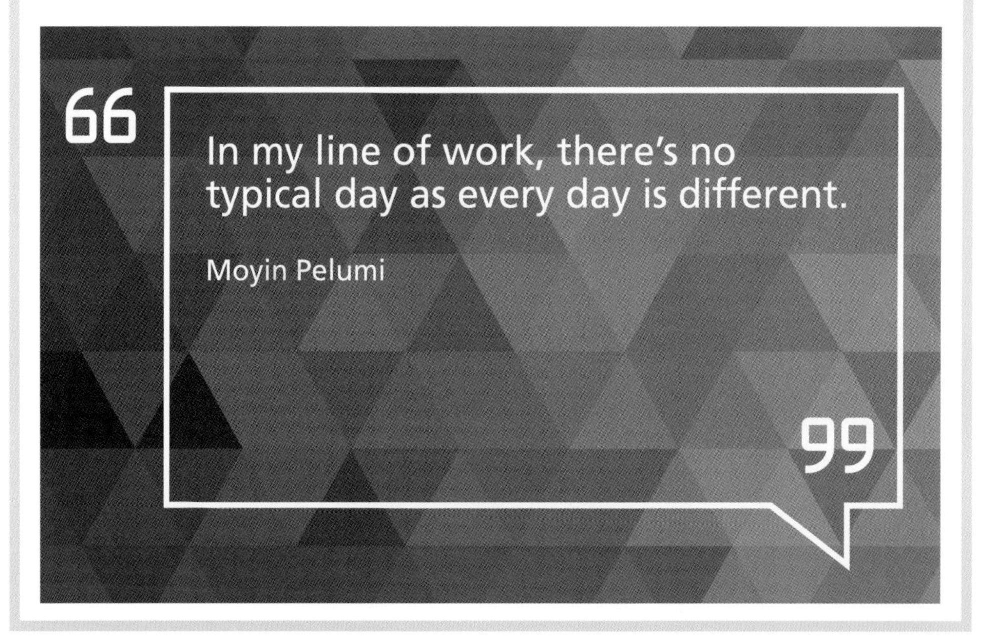
In my line of work, there's no typical day as every day is different.
Moyin Pelumi

increase the height.

This task was exceptionally fun for someone like me who loves challenges as we had limited Lego blocks and the aim of the challenge was to build to make profit. I was given the role of lead designer, so I was ecstatic when my team won the challenge with a height of just over a metre. It was a good team building experience.

Which individual project or person inspired you to become a civil engineer?

My choice to pursue a career in site engineering is a combination of a lot of factors.

During my A-levels, my favourite subjects were sociology, physics and maths so I guess already all directions were pointing towards engineering as I wanted to use my technical knowledge to improve society.

My first placement was with CH2M when I was 16 and we visited the Olympic Park in London. Although I enjoyed

my visit, I was sceptical about being an engineer because it involved a lot of design work. However, doing a second internship on site in Nigeria showed me a different side of engineering. I could be on site and liaise with different personnel to bring the project to life.

My mother also worked in the construction industry and her experiences inspired me to pursue a career in engineering.

I'm a civil engineer, but I'm also...

...an explorer who loves visiting different places and seeing other cultures.

What about being a Graduate Site Engineer gets you out of bed each morning?

The subcontractors phone call at 7 am definitely gets me out of bed. Joking aside, it's the thrill and the challenges that surround my project which motivate me to do what I do each and every day.

What's one great thing that you love about civil engineering that you didn't know until you started working in the industry?

The variation and interlinking disciplines and career paths are one thing that I did not initially know civil engineering offered but it is one thing I truly love about it.

Which civil engineering project (past or present) do you wish you'd worked on?

There are two major historical engineering projects I wished I had worked on:

- The Great Pyramids of Giza in Egypt are truly one of the greatest wonders of the world. The durability of the structure and the magnitude are truly a masterpiece considering the time it was built. With the tools used and the level of technology they had, it must have been amazing to watch.

- Park Güell in Barcelona also has a great blend between design architecture and engineering. The way nature, light and structure were brought together is spectacular. It must have been a very challenging yet fulfilling project to work on.

Name one civil engineering myth you'd like to bust

That there is a lack of diversity – my experience has been very positive in this regard and I feel that the industry is becoming more diverse with good opportunities for all.

What are the best projects you have worked on?

I'm currently based on the A303 on the structures team. This is a £135m scheme by Galliford Try for National Highways which will dual the current single carriageway section of road between Sparkford and Ilchester.

Working on the A303 has been exciting as it has allowed me to observe the construction of a new road, comprehend the underlying process, and learn about the approach used to build bridges.

What excites you about the future?

I anticipate the emerging role of new technology to be formidable in the construction industry. I am excited to see Galliford Try exploring technology further as I have witnessed first hand the benefits to health and safety, sustainability, building works and surveying (through drones). I am passionate that incorporating technology in the construction industry will help maximise efficiency and I look forward to seeing Galliford Try's role in this sphere.

Why do you love working at Galliford Try?

The graduate programme has built me a good foundation for my journey in construction. This has had a significant impact on the development of my soft skills, including how I communicate with both internal and external stakeholders. Galliford Try has also supported and facilitated my training and development with planning, design management and temporary works awareness courses which have significantly enhanced my skillsets from university. They've also supported me with my IEng training agreement, which has widened and deepened my understanding of the construction sector. All in all, it is an incredibly supportive environment which fosters learning and growth.

Next steps...

There are graduate opportunities in a wide range of sectors to suit all tastes across public, private and academic institutions. Public sector organisations typically involve local or regional authorities including councils but also combined authorities, subnational transport bodies or central government. Private sector organisations typically involve consultancies and transport operators (e.g. rail or bus companies) but also wider interest groups like financial and investment institutions or architect studios. All of these organisations vary in size and geographical scope so it's useful doing some research in terms of the workplace type that is best suited. Many of these often have dedicated graduate schemes that support the training, professional and personal development of early career professionals to find the areas of work most suitable (and enjoyable!) for everyone's individual skillsets.

Many civil engineering companies offer graduate schemes in transport infrastructure engineering. The structure of the graduate schemes varies between companies. Some schemes are very structured and graduate engineers will carry out similar work as other graduate colleagues for the duration of the scheme. Other schemes allow you to start working on real-world projects straight away to obtain knowledge and experience through on-the-job learning in a more self-driven programme.

02
Disciplines of Civil Engineering

Energy

Energy

Civil engineers play a vital role in designing, delivering and managing our energy infrastructure, and in working towards universal access to clean, affordable and reliable sources of energy.

Civil engineers working in the energy sector play an essential role in designing, building, operating and maintaining the essential infrastructure that's needed to ensure the supply of energy for heat, power and transport. The energy sector comprises a wide variety of activities and companies that are involved in research, discovery, production, transportation, and distribution of energy. It includes electricity generation, transmission and storage; the supply of oil and gas; nuclear power, and also other alternative sources of power such as hydroelectric, solar, wind and geothermal.

Within the power generation sector, a wide range of technologies are used to provide essential electricity to homes, businesses and industry. These include:

- **Coal, gas and diesel-fired power plants**
- **Wind power plants (offshore and onshore)**
- **Nuclear power plants**
- **Hydroelectric power plants, including dams, reservoirs and run-of-river hydropower**
- **Solar power plants**
- **Geothermal power plants**
- **Tidal power plants**

Careers in the energy sector may also focus on emerging technologies, such as carbon capture and storage, hydrogen and circular economy solutions, including bioenergy.

Renewable energy

Civil engineers that work in the renewable energy sector specialise in the planning, design, construction and maintenance of a variety of renewable energy technologies which includes hydropower, wind, solar and geothermal energy.

Hydropower is one of the largest sources of renewable energy and involves using the natural flow of moving water to generate electricity. The International Hydropower Association distinguish four main types of hydropower scheme: run-of-river, storage, pumped storage and offshore hydropower.*

*hydropower.org/iha/discover-types-of-hydropower

Hydropower projects are multi-disciplinary by nature, requiring collaboration between specialists in hydrology, hydraulics, geology, geotechnical, civil, structural, electrical and mechanical engineering, as well as environmental scientists (biologists, hydrologists, ecologists, and wildlife habitat specialists) who are able to assess environmental impacts and address environmental remediation. Projects in this area can involve design and construction of new schemes as well as upgrading existing dams for hydropower.

Offshore wind is a rapidly maturing technology that is poised to play an increasingly important role in future energy systems. The lifetime of an offshore wind farm can be greater than 30 years, from the initial design work to the final decommissioning, and work in this area may involve delivering projects at various stages of the development lifecycle, for example:

- **preliminary surveys, site selection and feasibility studies**
- **design of the wind farm**
- **geotechnical/foundation engineering**
- **substructure, superstructure and substation engineering**
- **installation and commissioning**
- **operations, maintenance, and decommissioning**

Both the volume and scale of renewable energy projects has increased significantly in recent years. While this has undoubtedly provided fantastic opportunities for those working in the sector it has also presented increased technical challenges. This is perhaps most evident in the offshore wind sector with the round 3 and 4

offshore wind farm projects increasing in size significantly, and moving further into deeper water. As a result foundation design has become increasingly complex with civil engineers now looking at XL mono-piles, jacket and gravity base foundations.

Alongside the more established renewable energy technologies there are developments in other areas such as tidal power, combined heat and power plants, and energy storage, which all offer further exciting opportunities for civil engineers.

Roles in this sector often involve:

- initial development and planning studies through to front-end design, project management and commissioning
- working in multi-disciplinary teams relating to energy, environmental, geotechnical, structural, mechanical, electrical and other specialist engineering disciplines
- combining renewable energy production with existing power systems.

Typical roles in this sector include:

- **Discipline engineer** – large numbers of discipline engineers are required to design and deliver these projects. These include civil and geotechnical engineers, structural engineers, electrical engineers, and materials specialists.
- **Project engineer** – project engineers are required to deliver procedures and processes for construction, installation and commissioning.
- **Project manager** – this is a key role in project delivery which involves managing work packages, and building and maintaining strong client relationships, as well as tracking delivery against defined margin, schedule and quality requirements, and raising safety awareness and ensuring engineering risk assessments are carried out.
- **Engineering manager** – managing teams of engineers, providing leadership, supporting and mentoring junior colleagues, and ensuring quality is maintained.
- **Planner** – some of these projects are very large, with potentially a lot of repetition, so good logistics and planning are absolutely essential to project success.

A career in renewable energy can be rewarding in a number of ways:

- The development of renewable energy is vital in helping to reduce carbon emissions, mitigating the risk of climate change and ensuring that carbon reduction targets of 2050 are met.
- There is an increasing variety of renewable energy technologies, some are well established, others are at the cutting edge of innovation, giving the opportunity to research and develop new energy generation methods.
- Many are drawn to this area by the opportunity to work on a broad range of technical challenges – for example, in the offshore wind sector there are challenges of optimising wind farm designs, increasing turbine size, working in more remote locations, and reducing costs.
- A rapidly growing sector, there is a range of opportunities for people looking for well-paid, long-term jobs that make a difference.

Ghazi-Barotha hydropower project – Ghazi-Barotha, Pakistan

The Ghazi-Barotha hydropower project is a hydroelectric scheme on the Indus River about 10 km west of the city of Attock in the district of Punjab, Pakistan. With five generators, the plant has a maximum capacity of 290 MW.

Ghazi-Barotha is a run-of-river hydroelectric plant. A run-of-river scheme is one with little or no water storage.

It relies either on the strength of flow of a river, or water from a reservoir or dam upstream.

At the time, the $2.25 bn scheme was one of the largest hydropower projects in the world.

Apart from funding from Pakistan itself, financial backing came from the World Bank and the Asian Development Bank, as well as European and Japanese banks.

The scheme diverts water from a 2.5 km-long barrage on the Indus River near the town of Ghazi, about 7 km downstream of the Tarbela Dam and its hydropower scheme.

The primary purpose of the Ghazi-Barotha project is to provide constant peak power at times when Tarbela is generating low amounts of electricity.

Water from Ghazi passes along a 52 km-long, 100 m-wide, concrete-lined

open channel to the power station at Barotha. Having gone through the scheme's five generators, it's returned to the river.

The project's acknowledged as being socially responsible – the barrage site and route of the channel were chosen to avoid existing villages where possible.

The scheme won the UK Energy Institute's International Platinum Award in 2006 for its innovative approach to providing clean and sustainable low-cost energy.

Difference the project has made

The Ghazi-Barotha scheme provides low-cost electricity for the Punjab district and other areas of Pakistan.

The plant generates 10% of Pakistan's power – making it a major contributor to the country's expansion programme.

Spoil banks created by the scheme were topped with fertile soil to create farmland. A spoil bank is a pile of rocks, earth or other waste created by excavation works.

How the work was done

The Ghazi-Barotha hydropower project saw engineers build a barrage close to the town of Ghazi on the Indus River.

The Ghazi barrage is 2.5 km long and 25 m high. Engineers used concrete and earth to create the structure.

A barrage is a type of diversion dam with gates that can be opened or closed to control how much water passes through them. Barrages are often used to regulate river flows for use in irrigation and other schemes.

The project team excavated a 52 km-long, 100 m-wide canal to carry the water to the scheme's power station at Barotha. Workers also laid five pressure pipelines measuring 10.6 m in diameter to carry water to the project's turbines.

Engineers laid a 225 km-long, 500 KV transmission line to carry power generated by the scheme. Other works included the construction of a new electricity substation and the extension of two existing substations.

Ghazi Barotha Hydropower Project:

youtube.com/watch?v=Vk5rX5ykbqI

CAREER PROFILE

Andrew Ndungu GMICE
Experienced Civil Engineer, SMEC International PTY Limited

I work for SMEC in Nairobi, Kenya, as part of a multi-disciplinary team to develop designs for dams and hydropower systems.

A highly versatile career

I was inspired to become a civil engineer because I was fascinated by tall buildings, and wanted to know how they work. I was good at physics in high school and this inspired me to pursue a degree in civil engineering. After completing the degree, I gained an interest in water systems and developed my skills around that.

I love being outdoors. Most of the projects I work on are in remote areas of Africa and have an on-site assessment visit. I get to see how the natural world is and appreciate first hand how our decisions impact the natural surroundings. These have been the highlight of my career so far.

As a civil engineer, you get to sit behind a desk if you want to. You get to be outdoors if you want to. You get to travel from site to site if you want to, perhaps even in different countries or continents. It's a highly versatile career.

My civil engineering experience

I carry out the design and reporting, as well as coordinate, manage and guide other staff working towards the delivery of projects. I also provide backstop support for other water infrastructure projects, such as water supply and sanitation projects.

At the moment we are working on improvement works for a dam. My task is to conduct the slope stability and seepage analyses and report the results and computations to the Lead Designer for review. At the next stage, I take it for detailed drawings, which are conducted by a draftsman under my guidance. All this will be tied to a design report and works specifications. In addition, I spend my time managing other aspects of the design.

The most interesting project I have ever worked on was a penstock inspection project in Rwanda. We were tasked with inspecting a penstock pipe 1800 mm in diameter and 250 m long. The terrain we had to work on went to about 70° gradient at some sections. I got to experience the dynamic forces in play following a water hammer effect from an induced load rejection on the power plant. The effect was equivalent to a mini earthquake.

> you get to sit behind a desk if you want to. You get to be outdoors if you want to. You get to travel from site to site... perhaps even in different countries or continents. It's a highly versatile career.
>
> Andrew Ndungu

A chance to curb climate change

Climate change is a real issue that our generation is experiencing, and also one that we have a chance to curb. In the industry, project financiers use policy instruments, or sustainability bonds to prompt developers to consider the climate change impacts that the project will have.

Addressing the impact of projects on climate change is important for all of us.

Opening up a larger pool of opportunities

I am a Graduate member of ICE, working towards becoming a Chartered Engineer. I want to be professionally recognised, and becoming chartered will show my ability and competency. Clients are confident when a professionally qualified engineer is involved.

ICE is a globally recognised body. Most local professional bodies in Africa know the institution, and recognise their capacity to execute their mandate. This adds value to the individual member and opens up a larger pool of opportunities in different locations around the world.

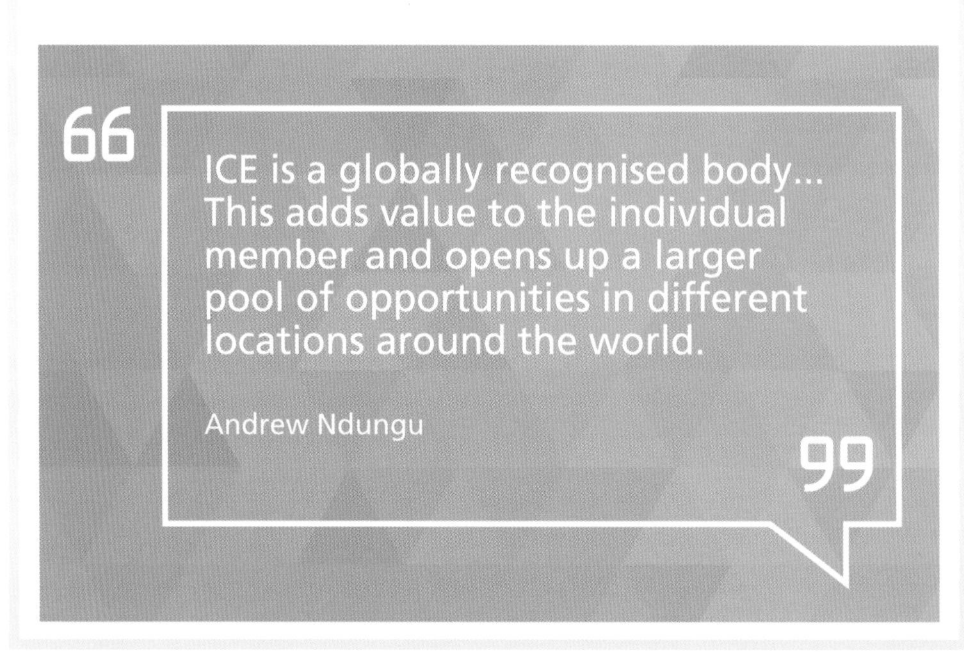
ICE is a globally recognised body... This adds value to the individual member and opens up a larger pool of opportunities in different locations around the world.
Andrew Ndungu

Nuclear

Careers in nuclear provide some of the most exciting and complex engineering challenges, requiring sound engineering judgement, teamwork and project delivery excellence.

Civil and structural works are a critical element of nuclear plants, and civil engineers in this sector play a vital role in the design, construction, operation, maintenance and decommissioning of a variety of nuclear structures. Nuclear structures include reactors, nuclear fuel element production installations, and structures such as fuel storage ponds as well as vitrification, encapsulation and radioactive waste storage facilities.

Civil works undertaken at nuclear sites include:

- heavy structural steel fabrication and erection
- significant reinforced concrete works
- land remediation and clearance
- protective works against extreme environmental events including floods and earthquakes
- transport systems for road and rail
- drainage and water treatment works
- electrical distribution and utilities
- foundation and backfilling works.

There are unique aspects to the nuclear industry including regulation, technologies, culture and terminology which mean there are particular construction requirements that differ from those involved in other areas of construction. These are also significant multi-disciplinary projects requiring a large variety of technical specialists from many fields.

Nuclear decommissioning is the process by which nuclear materials are removed from a structure and then the nuclear facility is dismantled, often requiring the construction of new structures and equipment, such that it no longer presents a potential hazard. It includes a range of processes and technologies such as dismantling, decontamination, and radioactive waste management, as well as environmental remediation, site clearance and reuse.

PROJECTS

Sellafield

Creating a clean and safe environment for future generations at Sellafield

Sellafield has been more than seven decades in the making, starting out as a munitions factory during the Second World War, followed by its central role in the UK atomic weapons programme in the 1950s, to the more recent nuclear reprocessing and storage facilities.

We now face a complex nuclear clean-up challenge for which there are no blueprints.

Our engineers at Sellafield Ltd face some of the most unique and challenging work in the industry, and our engineering capability enables us to accelerate high hazard reduction in a safe, environmentally responsible, cost effective and pragmatic manner.

From emergent, on-plant situations requiring support and advice, plant modifications on legacy facilities and infrastructure, to large multi-million pound new build projects, Engineering at Sellafield Ltd covers a great spectrum of opportunities for graduates.

Each of our infrastructure projects delivers vital work to create a clean and

safe environment for future generations.

The SIXEP Continuity Plant (SCP) is an example of this. When finished, the facility will help ensure the continued operation of our Site Ion Exchange Effluent Plant (SIXEP).

Operational since 1985, the existing Site Ion Exchange Effluent Plant (SIXEP), removes radioactivity from various site effluent streams on the Sellafield site and is coming to the end of its design life. It acts as the 'kidneys' of the site, filtering out radioactive material before discharging to sea.

The new facility, currently under construction and set to finish in 2029, will enable us to continue safely discharging to sea for decades to come and is needed until around 2060.

SCP is being delivered by the Programme and Project Partners (PPP), which is made up of lot partners: KBR (integration partner), Jacobs (design and engineering partner), Morgan Sindall Infrastructure (civils construction management partner), Doosan Babcock Ltd (process construction management partner) and Sellafield Ltd as the fifth partner.

Major milestones that have already been achieved are:

- completion of Detailed Design with the project described as best in class amongst other Sellafield projects
- business case approved by HM Treasury in December 2021, 3 months ahead of schedule

- building foundations completed ahead of schedule.
- a dozen contracts awarded to supply bespoke, specialised mechanical and electrical equipment to operate the facility
- three tower cranes installed to support construction of the structural elements
- 5 000 m^3 of concrete poured and first phase structural steelwork erected

As an engineering graduate, your technical ability and knowledge will enable you to support the delivery of these large projects. You will exhibit the correct behaviours, have a questioning attitude, and gain a good understanding of the project lifecycles from studies/concept design through to operation and decommissioning.

Over two years, design graduates are seconded to operating units throughout the business, enabling you to gain experience in a variety of aspects of engineering. Working towards becoming responsible for ensuring delivery to the correct quality and timescales.

As part of the graduate scheme, you will become involved with STEM activities, both internal and external to the organisation. Opportunities are available outside of your direct project delivery, such as becoming a member of the graduate council or a Centre of Expertise.

CAREER PROFILE

Josh Graves CEng MICE
Project Manager Pile 1 F&I Studies, Sellafield Ltd

How would you describe your current role? What do you enjoy about it?

I have recently started a new role in remediation at Sellafield, so I have a lot of training to complete and new skills to develop. I am enjoying the challenge of learning a new role. I am lucky in that I am joining a new project in its study phase which gives me the chance to gain the in-depth knowledge and develop my understanding along with the rest of the team.

What stands out as the most interesting project you've worked on?

My current role definitely has the potential to become the most interesting. I am working on the Pile 1, Fuel and Isotopes Study, an important step towards de-fuelling and making safe one of the most significant reactors in the UK's nuclear history. It is exciting to be involved in a project with this level of national importance.

Career development, education, first role, and career highlights

After completing GCSEs and A–levels I undertook a degree in civil engineering at the University of Salford.

After working in the Geotechnical and Rail sectors I was ready for a move back to my home county of Cumbria and in 2014 I was able to secure a job at Sellafield Ltd as a civil system engineer. In this role I was involved in asset management activities, initiating and overseeing refurbishment and maintenance projects for a range of nuclear facilities. I then moved on to a construction role (within Sellafield Ltd) as a section engineer in 2020, working on utilities projects such as bridge refurbishments. Most recently I have moved into a project management position where I am starting the development of new skills.

A personal highlight for me is completing my successful professional review in 2019 to become a Chartered Civil Engineer, an achievement I remain very proud of.

What changes in the profession have you seen and where do you think the profession is heading?

The obvious change during my career in civil engineering is the emphasis placed on sustainability and reducing our carbon footprint. Being part of the nuclear industry my role involves dealing with the legacy issues from decisions made in the past so considering the long-term implications of my decisions is something that is always in the front of my mind. Civil engineers have the influence to bring about big changes in how we construct and manage assets with the potential to make a huge contribution to sustainability. This is of particular importance in the nuclear industry where the current programme of remediation and decommissioning activities has the potential to reduce drastically the environmental risk from nuclear materials.

“A personal highlight for me is completing my successful professional review in 2019 to become a Chartered Civil Engineer, an achievement I remain very proud of.”

Josh Graves

Next steps...

Organisations in the energy sector range from large multinational organisations to smaller firms focusing on the development of specialist technology and services. The largest organisations use a range of fuel sources (oil, gas, coal, nuclear and renewables) to generate electricity for distribution to homes and industry. Most now operate nationally and even internationally. Smaller companies may focus on one locality or on one source of energy.

For graduate careers in renewable energy, in addition to general graduate recruitment websites, there are also sector specific websites such as The GreenJobs Network which includes a number of job boards relevant to the energy sector, including solarjobs, renewable energy jobs, and windjobs.

The nuclear sector has a useful website at nucleargraduates.com with further information on careers in this sector. Worldwide there are around 55 nuclear power plants under construction in 15 countries, and in the UK, construction of Hinkley Point C started in 2018 and is due for completion in 2027, and Sizewell C has recently been granted development consent. Further information can be found on the EDF Energy website.

Within the field of nuclear decommissioning, the Sellafield Ltd graduate and apprentice programme is based on three key elements of professional development:

- Specialist technical knowledge and professional development
- Understanding of the business
- Behavioural skills

02 Disciplines of Civil Engineering

Development, planning and urban engineering

Development, planning and urban engineering

Roles in development, planning and urban engineering focus on the early phases of planning infrastructure development and with the design of towns, cities and communities throughout the world. Projects can range from an expedient repair, reconstruction or repurposing, to an entirely new development. International development work focuses on the unique challenges of delivering vital services and infrastructure in developing regions.

Careers in planning and development seek to envision the future of towns, cities, regions and nations. Good planning seeks to address the economic, social, physical and environmental aspects of any development in a fully integrated manner. It involves reaching agreement on the purpose and scope of the venture, the nature

of the services the scheme is intended to deliver, and the benefits it is desirable to provide. It also requires planning and appraising future development in a manner that is sensitive to the risks, uncertainties and complexities of different contexts – that is, to ensure sustainability and resilience.

Development, planning and urban engineering spans many disciplines and civil engineering specialisations, and overlaps with career options covered in previous sections of this guide. It covers planning for transport facilities, including road, rail, air and seaborne transportation; planning for the development of energy infrastructure, such as power plants and renewable energy facilities, natural gas pipelines and heat networks; planning sewer and water services; and planning public facilities, such as hospitals, schools, universities, theatres, government buildings, shopping centres, parks and recreational areas.

Urban engineering

Urban engineering relates to the planning, design and construction of the infrastructure of towns and cities. It covers a broad range of different types of scheme, including highway planning (roads, footpaths, bike routes, parking facilities, street lighting); waste (sewer systems, solid waste management and materials storage); public facilities (schools, hospitals, parks and other shared spaces); residential housing, and industrial and commercial facilities. Work in this area will typically involve elements of urban planning and urban design.

Urban planning involves formulating plans for the development and management of urban and suburban areas. Some key areas of work include: helping to shape the long-term strategy for an area; conducting research into the feasibility of new developments; researching and assessing what the impacts of a project would be on the locality; managing the submission of planning applications, monitoring progress, and providing further advice and guidance post submission.

Urban design has been described as the art of making places. It is concerned with the arrangement, appearance and function of suburbs, towns and cities, and with the integrated design of buildings, public spaces, streets, transport and landscape. It can range from the design of transport and infrastructure networks, city centre development and regeneration, through to the micro scale of street furniture and lighting. Professionals in this area seek to understand urban spaces and how the streets, buildings, and the spaces in between can be arranged to make aesthetically attractive, economically vibrant, and environmentally sustainable places. An interdisciplinary area, urban designers work with professionals across a range of sectors, including specialists in planning, architecture and landscape architecture, economics, law and finance.

Projects in this area will often involve:

- evaluating impacts of broader trends such as urbanisation, climate change, and technology, and considering how towns and cities can become more resilient to a changing climate, how to house a growing and shifting population, and responding to technological advances
- long-term strategic planning for the expansion of cities and for programmes of urban renewal
- formulating plans dealing with land use in cities, towns, and counties which determine how to use land for homes, businesses and recreation
- designing entire communities, including subdivisions and neighbourhoods
- planning for the modernisation of transportation systems.

Roles in this area typically involve:

- predicting and modelling the future needs of a community
- gathering and analysing economic and environmental studies, censuses, and market research data, and conducting field investigations to analyse factors affecting land use
- providing strategic advice on planning proposals, managing consultation, assessing impacts, and preparing applications for planning consent
- providing urban design and masterplan advice for a range of different types of scheme
- meeting public officials, developers, and the public regarding development plans and land use.

Key skills include:

- an ability to solve complex problems and capitalise on opportunities
- keeping track of an ever-changing landscape and adapting to changes in social, political, economic, environmental and technological context
- excellent written, verbal and graphic communication, and for a range of different audiences
- strong research skills, including the ability to find information, analyse it, and make accurate conclusions based on the information
- proficiency in using tools for mapping and analysis of spatial information, including the use of urban design and planning software.

Future cities – Songdo: South Korea

Most of the world's cities have three things in common. They were built in places that were easy to defend, had access to fresh water and were convenient for trading.

As cities have evolved what they need to flourish has also changed – both in scale and complexity. Modern cities need reliable energy sources, efficient transport and effective sewerage and communications networks.

The challenge for civil engineers and planners for the future is to provide the infrastructure and services modern cities need while still providing a pleasant place to live.

The world's population is predicted to reach 8 billion by 2023, with most growth in developing countries such as Brazil, India and Pakistan. It's estimated that 70% of people will live in cities by 2050.

One way of meeting this predicted growth is to build cities from scratch. This is already happening in South Korea with the construction of Songdo – the world's first 'smart' city. The project has been designed around new technologies.

Like Songdo, urban planning from the foundations up isn't always possible. A major future challenge for civil engineers will be finding ways to retrofit existing infrastructures to meet the needs of city residents.

Any city that wants to thrive in the 21st century and beyond will need to adapt.

Work on technologies that could be used in future cities has already seen innovations and an idea of how some of tomorrow's urban infrastructures could look.

Examples include vacuum waste disposal. These are systems which transport waste along pneumatic tubes to collection stations where it's compacted and sealed in containers. This helps separation and recycling of waste.

Vacuum systems are already in use in China, South Korea and the US. Planned schemes include one for the city centre of Bergen in Norway.

Songdo in South Korea is generally credited as the world's first future city – buildings have automatic climate control and computerised access. Water, waste and electricity systems track and respond to the movement of residents.

But some critics have voiced fears that a computerised city like Songdo could also develop into a 'Big Brother' environment where everyone is under surveillance.

The challenge for engineers and planners will be to remember that cities are much more than nuts, bolts and hi-tech innovation. Urban infrastructure should serve the needs of citizens and communities – helping them to prosper.

An alternative for cities of the future may be using new technologies to integrate with existing systems, as well as putting the needs of citizens first when upgrading their environment.

Examples of this approach could include using smart sensors to measure water leaks in real time, control traffic flow (smart motorways) or dim street lights when no-one is around.

Yan Zhou
CEng FICE MHKIE CMgr FCMI
Technical Innovation and Digital Disruption Lead, Jacobs

Inspired to shape a better world

I read the story of Zhan Tianyou when I was a child. Zhan was the father of Chinese railroads who built the first major railway projects in China about 150 years ago. I was particularly impressed by his innovative railway design to overcome the problem caused by a steep mountain. It was the 'wow moment' when I thought civil engineers could change people's lives.

Going on to achieve ICE Fellowship at the age of 37 was probably the proudest moment of my career. I was told that I was one of the youngest Fellows in the world at the time.

The professional bedrock

Becoming a Chartered Engineer with ICE was the bedrock of my career. I was given more responsibility after becoming chartered and it provided me with confidence in dealing with the challenges I faced.

I received strong support from ICE, not only from the technical and professional development perspective, but also on broadening my network and connecting me with others. After I got chartered, I started volunteering for ICE by joining the London regional committee and now sitting on the ICE Council representing the London region. It gives me the platform to speak out for members in the region and help them to develop their career.

The importance of communication

Once I started working, I realised that good communication is the key to success. We need to communicate to the general public so they understand the benefits the infrastructure will bring to their life; we communicate to politicians and clients, so they are informed and professionally advised in their decision making process; we communicate with colleagues and co-workers so we can get the job done. Communication is one of the great things I love about civil engineering.

Facilitating innovation

My role in Jacobs' transportation division helps to facilitate technical innovation and to develop new ways of working across our business. This involves bringing together an array of key stakeholders to work towards a common vision and ensure our work is fully thought through and challenged.

Helping to achieve project efficiency and productivity provides great satisfaction.

The most interesting project I worked on was the Houses of Parliament Restoration and Renewal Programme. I was responsible for setting up the corporate services function of the programme, such as governance, finance, accounting, assurances, and HR procedures. This is not what you would typically expect to do but it gave me self-belief and showed that civil engineers have highly transferable skills.

> Becoming a Chartered Engineer with ICE was the bedrock of my career.
>
> Yan Zhou

Rebecca Houlder

Graduate Asset Management Consultant, Jacobs

What does asset management involve?

I work with a team looking at how our clients' existing assets – anything from a tunnel to a bridge, or even a lightbulb – are working. We assess their performance and identify if they need any improvements to continue working safely. No two days are the same; one minute I might be preparing for a presentation or client meeting, the next I'm building a carbon costing tool to calculate and compare the carbon impacts of solutions.

What do you like about your job?

I get to talk to my clients a lot, which I really enjoy. I also get a lot of responsibility and opportunity to put forward new ideas, which is something I didn't necessarily expect as a graduate. But my team is always on hand to provide support if I need it.

What inspired you to become an engineer?

Maths was always my favourite subject at school, so I thought civil engineering seemed like an interesting option. When I started looking into it, I realised there were so many different paths to follow – it's not only about being out on a construction site.

Would you recommend a career in civil engineering?

Absolutely, there really is something for everyone! You can work in design, be out on site, or in an office like me. There are also

many opportunities to travel or relocate within the industry. I would love to travel to Australia for work in the future!

How is the industry changing?

There's a huge drive right now to innovate to ensure everything we do lasts longer and has a positive impact on the climate. I'm always looking for new low carbon materials, technologies or methods to make our assets more sustainable.

Is there a project you wish you had worked on?

It would definitely have to be the Falkirk Wheel in Scotland. It's such a simple idea and much more efficient than a series of lock gates. I also think the Hoover Dam in the USA is really impressive!

Name one civil engineering myth you want to bust

People sometimes have a problem understanding what engineering – or consulting – exactly is, because no two careers will look the same, and this can often put people off applying. I want to help people understand the options they have in engineering so that more people, especially more women, want to come into the profession.

What advice would you give to people looking to get into engineering?

Take your time getting a good understanding of what engineering is, and the different roles available within the industry. Attending careers fairs and talking to people who actually work as engineers will be helpful. The 'Who are civil engineers?' section on the ICE website is a great place to start because you can easily see the variety of job titles people have!

Do you have any hobbies?

I love playing sports like tennis, skiing and snowboarding. I worked as a chalet host in St Anton, Austria before I joined Jacobs in 2021.

International development

International development work focuses on the application of civil engineering in humanitarian contexts, helping to increase access to basic services and supporting poverty alleviation and good governance. International development projects seek to provide appropriate levels of healthcare, education, infrastructure, water, sanitation and environmental protection. The infrastructure problems presented by the developing world are among the most challenging the civil engineering sector faces, and the help civil engineers can offer is invaluable.

Within this area, engineers also play a vital role in helping communities respond to natural disasters such as floods, earthquakes, and storms, and in post-conflict reconstruction and development. Projects can range from reinstating urban power and water systems to the construction of multi-storey buildings and bridges. In most humanitarian situations the framework in which engineering takes place will be very different to most engineers' experience of professional practice. It is particularly important for those working on these projects to be aware of the wider political, financial, environmental, cultural and social context in which the projects are taking place.

Tina Gunnarsson
CEng MICE
Senior Methods Engineer, Balfour Beatty

Championing sustainability

At school, I fundraised for the charity Tabitha and travelled to Cambodia to build 15 houses for a small village so their houses wouldn't get washed away during the monsoon season. I discovered the large-scale positive impact that engineers could have on communities and improving future sustainability.

What I enjoy most is the diversity of roles within civil engineering. You can be on your feet all day on site, doing calculations in the office, creating 3D models, writing bids, working with legal or commercial, championing sustainability or safety.

I'm currently a Senior Methods Engineer, developing low carbon construction methods as we work towards zero waste and beyond net-zero carbon by 2040. Four years ago, my typical day involved being outside on site constructing a 180 m-tall tower. Three years ago I worked in a design office, creating technical drawings and doing temporary works calculations. In my current role, I'm learning and researching low carbon methods and sharing that knowledge to help engineers make sustainable decisions.

Benefitting local communities

My career highlight and favourite project to date is Minigo Bridge in Rwanda. I was part of a team of ten that fundraised and constructed a 36 m span suspension footbridge for a village in the Rubavu region with the charity Bridges to Prosperity. It was a lot of fun to build physically, use power tools and climb scaffolding.

The scheme didn't break any records – it wasn't the biggest, most incredible engineering feat or longest span – but it's the only project where I've met or played games with a large portion of the 800 children that it would benefit.

The experience changed me on a personal level. I returned from the trip as a vegetarian after befriending the dinner goat and started making big changes to try to live more sustainably.

Improving diversity

I co-founded SheBuilds Collective, a community for women in construction, and I co-host the Engineering Rebuilt podcast. I am incredibly fortunate to be part of a great group of supportive and empowering women, and working with them is often a highlight in my day.

I would like to see an accelerated change in construction culture, to become more inclusive and diverse. I'd like to see women staying and having successful, varied and fulfilling careers because it's become a more equal place to work, and for those joining the industry not to experience discrimination while at work. The more diverse we are, the more creative thinking and new solutions we can implement as an industry.

Becoming a better engineer

My main motivation for becoming professionally qualified as a Chartered Engineer with ICE (CEng MICE) was to gain confidence in my engineering capabilities. To have my knowledge and abilities independently verified by ICE provided me with an assurance that I could make good engineering decisions.

Going through the process to become chartered on the ICE Training Scheme provided me with a lot of structure as a graduate. This helped me to develop a breadth of knowledge I otherwise wouldn't have and reinforced my engineering foundations. It was an invaluable experience that allowed me to build a portfolio and reflect on my experiences, making me a better engineer.

Being professionally qualified, as well as being awarded the James Rennie Medal in 2020 for best chartered professional review candidate, has allowed me to take on greater roles of responsibility in my career.

PROJECTS

Rural Access Programme (RAP)3, Nepal

Connect remote communities through a new transport infrastructure

Poor road infrastructure traps some 80% of Nepal's population in subsistence agriculture and makes transport treacherous, which shoots up prices of basic commodities. The award-winning, UK Aid-funded Rural Access Programme, now in its third phase (RAP3), uses the construction of transport infrastructure as an entry point for improving the lives of the poorest people in Nepal's remote areas.

RAP3 has built 100 kilometres and maintains over 2 000 kilometres of climate-resilient roads that connect remote communities to markets, healthcare and education facilities. It employs 9 000 poor people, 40% of whom are women, in road construction and maintenance.

The programme also builds partnerships between banks, multinationals, micro, small and medium enterprises (MSMEs) and farmers to ensure that the poor benefit from improved access.

RAP3 has helped over 3 000 subsistence farmers to enter long-term agreements with businesses and shift to commercial agriculture. The partnership with Unilever has resulted in the appointment of 228 female rural sales agents, which provide remote communities with essential products. MSMEs partners are introduced and connected with national retailers and international importers. RAP3 has attracted £518,000 in private sector investment at the rate of £2.50 for each pound of UK aid invested.

Project elements

Sustainability is embedded in RAP, as shown by the programme's holistic approach to improving the long-term livelihoods of the communities involved in this project.

Design standards for roads include climate change adaptation and disaster resilience features. RAP's roads are built using environmentally sound methods which, together with maintenance, allow roads to be dependable in all weather conditions. These include the use of bioengineering to stabilise slopes rather than engineered structures.

RAP has introduced a culture of road maintenance in Nepal where 'build and forget' had become normal practice in all national district road programmes. By 2012, over half of the local road network constructed over the past 15 years was no longer useable because of lack of maintenance. The government of Nepal decided to launch and pay for road maintenance groups (RMGs) throughout most of the country, drawn from the poorest households.

RMGs operate throughout the year, so they can prevent much of the damage caused by the monsoons. They generally work half-time, which allows them to pursue other activities, such as farming, livestock rearing and child-rearing. Importantly, the government's investment in roads now amounts to the equivalent of £1.5 million per year.

RAP construction and maintenance teams are embedded in technical offices responsible for engineering works in the districts. This enabled the programme to respond quickly to the 2015 earthquake-affected districts in which RAP operates, clearing local roads to allow access to the first responders such as the army and emergency services.

RAP deliberately uses labour-intensive methods which generate employment for local people, including women, many of whom are earning a wage for the first time. This is improving their status in society. Moreover, the programme is promoting gender parity by appointing women as community leaders in villages and raising the profile of International Women's Day in Nepal.

To help address the skills gap in the country's engineering industry, RAP runs a six-month internship scheme that allows young people to gain hands-on experience in engineering and socioeconomic development under the mentorship of experienced professionals. After the internship, participants are often then absorbed into the RAP graduate programme. Nearly 120 interns and currently 30 graduate engineers have participated so far.

Dr Priti Parikh CEng FICE
Professor, UCL and Head, Engineering for International Development Centre

Transforming living conditions

At UCL, I head the Engineering for International Development (EFID) Research Centre in the Bartlett School of Sustainable Construction. EFID researches infrastructure solutions (water, sanitation and energy) for resource-challenged settings globally.

Our vision is that every person on this planet should have access to affordable, appropriate and inclusive infrastructure. We provide evidence through research to policy makers and practitioners to inform decision making on infrastructure, taking into consideration the UN Sustainable Development Goals and climate change.

Communities need access to basic infrastructure services to improve their quality of life and build resilience. It is all well and good talking about climate change, but if a household does not have access to water and electricity, they will not be able to engage and respond to it.

My centre explores climate resilient infrastructure, with a focus on gender inclusive solutions as women bear the burden of poor infrastructure and climate change in marginalised communities.

Civil engineering is so rewarding. I have seen how infrastructure can transform living conditions in slums and villages around the world. It can lead to improvements in human settlements and have a knock-on impact on health, education, housing and incomes.

Making a difference and inspiring others

After studying structural engineering and urban planning in India, I worked in the industry for 12 years in India and the UK on infrastructure planning projects.

I then pursued a doctorate at the University of Cambridge. During my PhD I examined the impacts of infrastructure in slums in India and townships in South Africa. The doctorate made me realise that in addition to project delivery, I was interested in research. I have now been in academia for 12 years.

My industry experience has inspired me always to consider the application of research to real-world challenges and the need to engage actively with industry and policy makers to influence engineering thinking.

My most rewarding achievement is being recognised as one of the 100 most influential academics for policy making in sustainability and climate. I enjoy that I can make a difference as an individual and motivate my students and researchers to do the same.

In 2022, I was recognised as one of the top 50 women in engineering by the Women's Engineering Society.

The confidence to lead and deliver projects

As we deal with the big challenges of urbanisation, climate change and conflict we need engineers who can future proof, anticipate unexpected challenges and solve problems.

During my time in industry I was encouraged to pursue a professional qualification – Chartered Engineer in my case.

This gave me the confidence to say – yes I have the skillsets to lead and deliver projects. It also gave my employers and clients extra confidence to support my career progression.

A passionate community

I am proud to be a Fellow of ICE. ICE membership has opened up networks and provided access to valuable resources. I feel part of a community which is passionate about societal impact and environmental protection. When I walk into ICE's headquarters it is a reminder of what engineering as a profession can achieve.

Recently I was elected to the ICE Council where I am now learning more about the amazing work done by the institution to influence policy, outreach to get more people interested in engineering and engineering leadership.

The more I learn about ICE's activities, the more inspired I get – and wish that I had become a member much earlier on in my career.

Next steps...

There are graduate opportunities in a wide range of sectors across public, private and academic institutions. Public sector opportunities in this area include government policy researchers and analysts, junior planners at all levels of government, UNOPS and similar international organisations, such as the World Bank Group, ICRC, UNDP and the Red Cross. There's a growing preference for a postgraduate degree in a relevant subject, such as planning and development.

In the private sector there are graduate roles in specialist infrastructure consultancies, utilities, mining and oil and gas companies, and consultants serving these industries, as well as urban design consultancies, and international infrastructure consultancies.

Other suitable positions might be in academic or research institutions. Non-governmental organisations (NGOs), international non-governmental organisations (INGOs), UN agencies, and large private sector consultancies dominate the development and humanitarian sectors. They have very different structures and business models, which have a significant effect on what it is like to work for them.

03
Directory of Employers

Contact details of all ICE Corporate Partners and Approved Employers

Company	Website	Further information
ABA Consulting	aba-consulting.co.uk	aba-consulting.co.uk/contact-us
Abbey Pynford Group	abbeypynford.co.uk	abbeypynford.co.uk/contact
Aberdeenshire Council	aberdeenshire.gov.uk	askHR@aberdeenshire.gov.uk
ABS Group	abs-group.com	enquiriesuk@abs-group.com
ADNOC Onshore (Abu Dhabi Company for Onshore Petroleum Operations)	adnoc.ae/en/adnoc-onshore	jobs.adnoc.ae
Abu Dhabi Fund for Development	adfd.ae	adfd.ae/english/opportunities
Acies Civil and Structural Limited	aciesgroup.co.uk	aciesgroup.co.uk/contact-us
Adept Consulting Engineers Limited	adeptcsce.com	adeptcsce.com/careers
Adkins Consultants Ltd	adkinsconsultants.com	office@adkinsconsultants.com
AECOM	**aecom.com**	**aecom.jobs/civil-engineer/jobs-in**
Aggregate Industries	aggregate.com	careers.aggregate.com
AKT II (Akt II)	akt-uk.com	akt-uk.com/careers
Alan Baxter and Associates	alanbaxter.co.uk	personnel@alanbaxter.co.uk
Alan White Design Ltd	alanwhitedesign.com	alanwhitedesign.com/contact
Alan Wood & Partners	alanwood.co.uk	eng@alanwood.co.uk

Alan Wood & Partners

alanwood.co.uk
eng@alanwood.co.uk

Since 1968, Alan Wood & Partners has been a market leading provider of professional management and engineering design services, now with offices in six locations across Yorkshire and Lincolnshire.

With over 110 staff and supporting in the region of £1bn of projects annually, we have acquired a sound understanding of the development and construction sector. We thrive in this challenging and technically diverse environment, meeting the often conflicting pressures of managing sustainable development in sensitive environments.

We work with clients to realise their aspirations and plans from production of initial feasibility and concept studies through to delivery and operation.

Our personal approach to all our commissions has the same underlying philosophy – to work closely with our client and other team members, to develop the most creative engineering solution.

We do this by supporting our pool of over 90 technicians and engineers ensuring continued professional development through apprenticeships, degrees and charterships, as well as creating a workplace where talented people can thrive.

If you would like to know more about AWP and our career development programme please view **www.alanwood.co.uk**

Company	Website	Further information
Albert Fry Associates	albertfryassociates.com	albertfryassociates.com/careers
ALEC (ALEC Engineering & Contracting LLC)	alec.ae	alec.ae/opportunities
Alpha Construction Limited	alphaconstruction.co.uk	alphaconstruction.co.uk/careers
Alstom Ltd	alstom.com	alstom.com/careers
Alun Griffiths (Contractors) Ltd	griffiths.co.uk	griffiths.co.uk/jobs
AmcoGiffen	amcogiffen.co.uk	amcogiffen.co.uk/careers
Amey	**amey.co.uk**	**amey.co.uk/your-career/early-careers**
Andrew Scott Ltd	andrewscott.co.uk	andrewscott.co.uk/careers
Andun Engineering Consultants Ltd	andun.co.uk	andun.co.uk/careers
Anglian Water Services Ltd	anglianwater.co.uk	anglianwatercareers.co.uk
Apex Consulting Engineers	apexconsulting.co.uk	apexconsulting.co.uk/careers
Apollo Offshore Engineering Limited	apollo.engineer	apollo.engineer/careers
Aqua Consultants	aquaconsultants.com	aquaconsultants.com/careers
Aquaterra Energy Limited	aquaterraenergy.com	aquaterraenergy.com/careers
Arc Partnership	arc-partnership.co.uk	arc-partnership.co.uk/careers
Arcadis	arcadis.com	careers.arcadis.com
Argyll and Bute Council	argyll-bute.gov.uk	argyll-bute.gov.uk/jobs
Arup	arup.com	arup.com/careers
A-Squared Studio Engineers Ltd	a2-studio.com	a2-studio.com/careers
Associated British Ports	abports.co.uk	careers.abports.co.uk
Assystem	assystem.com/en	assystemstup.com/assystem-stup-career
Atkins	**atkinsglobal.com**	**atkinsglobal.com/careers**
Atomic Weapons Establishment (AWE)	awe.co.uk	awe.co.uk/careers
Aurecon Australia Pty Ltd	aurecongroup.com	aurecongroup.com/careers
Avie Consulting Ltd	avie-consulting.co.uk	avie-consulting.co.uk/careers
Avove Limited	avove.co.uk	avove.co.uk/careers
Awcock Ward Partnership (AWP)	awpexeter.com	awpexeter.com/about-us/recruitment
Ayrshire Roads Alliance	ayrshireroadsalliance.org	ayrshireroadsalliance.org/What-we-do
Babcock International Group: Devonport Royal Dockyard	babcockinternational.com/sustainability/social/community/devonport-royal-dockyard/	babcockinternational.com/careers
Babcock Rail	babcockinternational.com/what-we-do/land/engineering-and-training-services/rail/	babcockinternational.com/careers
Bachy Soletanche Limited	bacsol.co.uk	bacsol.co.uk/people/career-opportunities
BAE Systems	baesystems.com	baesystems.com/en/careers
Bailey Partnership (Consultants) LLP	baileypartnership.co.uk	baileypartnership.co.uk/careers

Company	Website	Further information
BakerHicks	bakerhicks.com	bakerhicks.com/en/careers
Balfour Beatty	**balfourbeatty.com**	**balfourbeattycareers.com**
BAM Nuttall	**bamnuttall.co.uk**	**bamnuttall.co.uk/careers**
Barhale	barhale.co.uk	barhale.co.uk/careers-new
Bauer Technologies Ltd	bauertech.co.uk	bauertech.co.uk/en/contact/career
BCM Construction Ltd	enable-infrastructure.com	enable-infrastructure.com/contact
Beam Consulting Engineers Ltd	beamconsulting.co.uk	beamconsulting.co.uk/careers
BEAR Scotland Limited	bearscot.com	bearscot.com/our-people/careers-at-bear-scotland
Bechtel	bechtel.com	jobs.bechtel.com
Beckett Rankine	beckettrankine.com	beckettrankine.com/careers
Belfast Harbour Commissioners	belfast-harbour.co.uk	belfast-harbour.co.uk/careers
Berkeley West Thames	berkeleygroup.co.uk	berkeleygroup.co.uk/about-us/careers
Betts Associates Limited	betts-associates.co.uk	betts-associates.co.uk/contact
BG&E Consulting Engineers Ltd	bgeeng.com	bgeeng.com/career
Binnies UK Ltd.	binnies.com	binnies.com/early-careers
Black & Veatch Limited	bv.com	bv.com/contact-us
Blyth & Blyth Consulting Engineers Limited	blythandblyth.co.uk	blythandblyth.co.uk/careers
Bolton Council	bolton.gov.uk	bolton.gov.uk/jobs-skills-training
Booth King Partnership Limited	booth-king.co.uk	booth-king.co.uk/careers
Boskalis Westminster Ltd	westminster.boskalis.com	westminster.boskalis.com/careers
Bournemouth, Christchurch & Poole Council	bcpcouncil.gov.uk	bcpcouncil.gov.uk/Jobs-and-apprenticeships
BP Azerbaijan	bp.com/en_az/azerbaijan/home.html	bp.com/en_az/azerbaijan/home/careers.html
Breheny Civil Engineering	breheny.co.uk	breheny.co.uk/job-vacancies
Bridge Civil Engineering Ltd	bridgecivileng.co.uk	bridgecivileng.co.uk/careers
Bridgend County Borough Council	bridgend.gov.uk	bridgend.gov.uk/my-council/jobs
Britcon (UK) Limited	britcon.co.uk	britcon.co.uk/working-for-us
Brookbanks Consulting Ltd	brookbanks.com	brookbanks.com/careers
Bryden Wood Ltd	brydenwood.com	brydenwood.com/careers
BSP Consulting	bsp-consulting.co.uk	bsp-consulting.co.uk/careers
BT Bell Consulting Engineers	btbell.co.uk	btbell.co.uk/careers
Buckingham Group Contracting Ltd	buckinghamgroup.co.uk	buckinghamgroup.co.uk/careers
Building Design Partnership (BDP)	bdp.co.uk	bdp.com/en/careers-at-bdp/vacancies
Built Environment Design Partnership Limited (BE Design)	bedesign-group.com	bedesign-group.com/contact
Buro Happold	burohappold.com	burohappold.com/careers
Burroughs	burroughs.co.uk	burroughs.co.uk/work-with-us

Company	Website	Further information
Burrows Graham Limited	burrowsgraham.com	burrowsgraham.com/vacancies
BUUK Infrastructure	bu-uk.co.uk	buuk.current-vacancies.com/Careers
BWB	bwbconsulting.com	bwbconsulting.com/discoveracareer
Byrne Bros. (Formwork) Ltd	byrne-bros.co.uk	byrne-bros.co.uk/careers
Byrne Looby Partners	byrnelooby.com	byrnelooby.com/careers
Cadarn Consulting Engineers Ltd.	cadarnconsulting.co.uk	cadarnconsulting.co.uk/careers
Cairn Cross Civil Engineering Ltd	cairncross.uk.com	cairncross.uk.com/careers
Calderdale Metropolitan Borough Council	calderdale.gov.uk	calderdale.gov.uk/v2/residents/jobs-and-volunteering
Caley Water Ltd	caleywater.co.uk	caleywater.co.uk/WorkingWithUs
Campbell of Doune Ltd	campbellofdoune.co.uk	campbellofdoune.co.uk/recruitment
Campbell Reith LLP	campbellreith.com	campbellreith.com/careers
Canal & River Trust	canalrivertrust.org.uk	canalrivertrust.org.uk/about-us/work-for-us
Canary Wharf Contractors Ltd	canarywharf.com	group.canarywharf.com/careers
Canterbury City Council	canterbury.gov.uk	canterbury.gov.uk/jobs-and-volunteering
Card Geotechnics Limited (CGL)	cgl-uk.com	cgl-uk.com/careers-engineering
Cardiff University	cardiff.ac.uk	cardiff.ac.uk/jobs
Carey Group	careys.co	careys.co/careers
Carmichael Site Services Ltd.	carmichaeluk.com	jobs.carmichaeluk.com
Cass Hayward LLP	casshayward.com	casshayward.com/about/working-for-us
Cathie Associates UK	cathie-associates.com	cathiegroup.com/careers
Caulmert Limited	caulmert.com	caulmert.com/careers
Caunton Engineering Ltd	caunton.co.uk	caunton.co.uk/CareersVacancies
Cavendish Nuclear Limited	cavendishnuclear.com	cavendishnuclear.com/careers
China Road and Bridge Corporation	crbc.com	crbc.com.hk/en/OurPeople/Careers
City of Bradford Metropolitan District Council	bradford.gov.uk	bradford.gov.uk/jobs/apply-for-a-council-job/bradford-council-vacancies
Civic Engineers Ltd	civicengineers.com	civicengineers.com/career
Clancy Consulting Limited	clancy.co.uk	clancy.co.uk/careers
Clarkebond (UK) Limited	clarkebond.co.uk	clarkebond.com/careers
Coffey Geotechnics Ltd	coffeygroup.com	coffeygeotechnics.co.uk/contact-us
Colas Ltd	colas.co.uk	colas.co.uk/careers
Colas Rail	colasrail.co.uk	colasrail.co.uk/join-us
Concentric Construction Limited (CRL)	manking.com.hk	concentric@ccl-concentric.com
Concrete Repairs Ltd	concrete-repairs.co.uk	crl.uk.com/join-the-team
Conisbee	conisbee.co.uk	conisbee.co.uk/practice/careers

Company	Website	Further information
Considine Limited	considine.co.uk	considine.co.uk/contact-us
Construction and Procurement Delivery (CPD)	finance-ni.gov.uk/ construction-procurement-delivery	finance-ni.gov.uk/contact
Construction Marine Limited (CML)	cml.uk.com	cml.uk.com/careers
Continental Engineering Corporation	continental-engineering.com	continental-engineering.com/join-cec
Cormac	cormacltd.co.uk	cormacltd.co.uk/work-for-us
Costain	**costain.com**	**costain.com/careers**
COWI UK Ltd	cowi.com	cowi.com/careers
Craddy Pitchers Ltd	craddypitchers.co.uk	craddys.co.uk/join-us
Create Consulting Engineers Ltd	createconsultingengineers.co.uk	createconsultingengineers.co.uk/About-Us/Job-Opportunities.ice
Crouch Waterfall	crouchwaterfall.co.uk	crouchwaterfall.co.uk/people-and-careers
Croudace Homes Limited	croudace.co.uk	careers.croudace.co.uk
Crucis Designs Limited	crucisdesigns.com	crucisdesigns.com/recruitment
CTP Consulting Engineers	ctp-llp.com	ctp-llp.com/careers
Cullen Grummitt & Roe (CGR Group)	cgrgroup.com	cgrgroup.com/page/careers

Company	Website	Further information
Cundall	cundall.com	cundall.com/careers
Curtins Consulting Limited	curtins.com	careers.curtins.com
CWIC EDS	cwic.wales	cwic@uwtsd.ac.uk
Datrys Consulting Engineers Ltd	datrys.net	datrys.net/recruitment
David Carr Consulting Engineers Ltd	No website	enq@davidcarrconsulting.co.uk
David R. Murray & Associates LLP	davidrmurray.co.uk	davidrmurray.co.uk/Contacts
Davies Maguire	dmag.com	dmag.com/recruitment
DeepOcean Group	deepoceangroup.com	deepoceangroup.com/people/early-careers
Department for Infrastructure NI	infrastructure-ni.gov.uk	info@infrastructure-ni.gov.uk
Derbyshire County Council	derbyshire.gov.uk	derbyshire.gov.uk/working-for-us/jobs/find-a-job-with-us
Design2e Limited	design2e.co.uk	design2e.co.uk/careers
Design4Structures	design4structures.com	design4structures.com/careers
Devon County Council	devon.gov.uk	devonjobs.gov.uk/jobs
DJ Goode & Associates Ltd	djgoode.co.uk	djgoode.co.uk/careers
DLT Engineering Ltd	dlteng.com	enquiries@dlteng.com
DNV Services UK Limited	dnvgl.com	dnv.com/careers/job-opportunities
Doran Consulting Limited	doran.co.uk	doran.co.uk/careers
Dorset Council	dorsetcouncil.gov.uk	dorsetcouncil.gov.uk/jobs-and-careers/jobs-and-careers
Dougall Baillie Associates	dougallbaillie.com	dougallbaillie.com/careers
Dragados SA	dragados.com	dragados.com/htmlEN/contact
Drax Power Limited	drax.com	drax.com/careers
Dudleys Consulting Engineers	dudleys.co.uk	info@dudleys.co.uk
Dundee City Council	dundeecity.gov.uk	dundeecity.gov.uk/service-area/corporate-services/human-resources-and-business-support/council-job-vacancies
Dwr Cymru (Welsh Water)	dwrcymru.com	jobs.dwrcymru.com
Dyer & Butler Ltd	dyerandbutler.co.uk	dyerandbutler.co.uk/contact-us
E&M West	eandmwest.co.uk	eandmwest.co.uk/#contact-top
East Riding of Yorkshire Council	eastriding.gov.uk	eastriding.gov.uk/council/council-jobs-and-pensions/jobs-at-the-council
Eastwood & Partners (Consulting Engineers) Ltd	eastwoodandpartners.com	eastwoodce.com/careers
Ecotech Consulting Engineers Ltd	ecotechengineers.co.uk	info@ecotechengineers.co.uk
Edenvale Young Associates Ltd	edenvaleyoung.com	info@edenvaleyoung.com
EDF Energy	edfenergy.com/about/nuclear	edfenergy.com/careers

Company	Website	Further information
Egis Industries UK	egis-group.com/locations/europe-central-asia/united-kingdom	egis-group.com/why-egis
Egniol	egniol.com	egniol.com/careers
Eiffage Génie Civil UK	www.eiffage.co.uk	eiffage.co.uk/home/contact
EirEng Consulting Engineers Ltd	eireng.com	eireng.com/careers
Elliott Wood Partnership Ltd	elliottwood.co.uk	elliottwood.co.uk/contact
Engenuiti	engenuiti.com	engenuiti.com/jobs
Environment Agency	**environment-agency.gov.uk**	**gov.uk/government/organisations/environment-agency/about/recruitment**
Eurovia UK Limited	eurovia.com	euroviacareers.co.uk/graduates
Expedition Engineering Ltd	expedition.uk.com	expedition.uk.com/careers
Face Consultants Ltd	face-consultants.com	info@face-consultants.com
Fairhurst Group LLP	fairhurst.co.uk	fairhurst.co.uk/careers
Farrans Construction Ltd	farrans.com	farrans.com/careers
FCC Construccion SA UK Branch	fccco.com	fccco.com/en/web/reino-unido/people/send-cv
Ferrovial Construction (UK) Ltd	ferrovial.com	ferrovial.com/en-gb/careers
FJD Consulting (& Design) Ltd	fjdconsulting.co.uk	fjdconsulting.co.uk/people
FLUOR LIMITED	fluor.com	fluor.com/careers
Foundation Piling Limited	foundation-piling.co.uk	foundation-piling.co.uk/contact
FP McCann Limited	fpmccann.co.uk	fpmccann.co.uk/careers
Frankham Consultancy Group	frankham.com	frankham.com/working-with-us
Fugro GB Marine Ltd	fugro.com	fugro.com/careers
Fulton Hogan Ltd	fultonhogan.com	fultonhogan.com/careers
Galliford Try plc	gallifordtry.co.uk	careers@gallifordtry.co.uk
Gammon Construction Limited	gammonconstruction.com	gammonconstruction.com/en/career
Gatwick Airport Ltd	gatwickairport.com	gatwickairport.com/business-community/careers/
Gavin & Doherty Geosolutions Ltd	gdgeo.com	gdgeo.com/careers
Geomarine Ltd	geomarine.gg	geomarine.gg/careers
Geoquip Marine	geoquip-marine.com	careers.geoquip-marine.com
George Leslie Ltd	georgeleslie.co.uk	georgeleslie.co.uk/careers
GH Bullard & Associates LLP	ghbullard.co.uk	info@ghbullard.co.uk
GHD (Gutteridge Haskins & Davey Ltd)	ghd.com	ghd.com/en/careers
Glanville Consultants Limited	glanvillegroup.com	glanvillegroup.com/careers
Golder Associates (HK) Limited	golder.com	golder.com/careers
Goodson Associates	goodsons.com	goodsons.com/contact/
Graham Construction	graham.co.uk	graham.co.uk/careers

Company	Website	Further information
GroundSolve Ltd	groundsolve.com	enquiries@groundsolve.com
Hampshire County Council	hants.gov.uk	hants.gov.uk/jobs/careers/engineeringcareers
HanmiGlobal	hmglobal.com/en	hmglobal.com/en/about/recruit
Harley Haddow Ltd	harleyhaddow.com	harleyhaddow.com/careers
Hartigan	hartigan.co.uk	hartigan.co.uk/work-with-us
Haydn Evan Consulting Ltd.	haydnevans.co.uk	haydnevans.co.uk/careers
HBPW Consulting LLP	hbpw.co.uk	mail@hbpwconsulting.co.uk
HDR	hdrinc.com	hdrinc.com/careers
Health & Safety Executive	hse.gov.uk	careers.hse.gov.uk
Heathrow Airport Limited	heathrow.com	heathrow.com/company/careers
HEB Contractors Ltd	hebcontractors.co.uk	hebcontractors.co.uk/careers
Hewson Consulting Engineers	hewson-consulting.com	hewson-consulting.com/join-us
Heyne Tillett Steel	hts.uk.com	hts.uk.com/careers
High Speed Two Limited	hs2.org.uk	hs2.org.uk/jobs-and-skills/careers-with-hs2-ltd
Hochtief (UK) Construction Ltd	hochtief.co.uk	hochtief.co.uk/contact
HOP Consulting Ltd	hop.uk.com	hop.uk.com/careers

Bentley

We are JN Bentley, a leading civil engineering company delivering a varied portfolio of exciting and innovative projects across the UK.

Part of the Mott MacDonald Group, we employ 1400 talented people in a host of roles that means we can offer our clients all aspects of civil engineering services – from feasibility to design to construction and commissioning.

Celebrating our 50th birthday in 2022, we're proud of our Yorkshire roots but can now boast premises and sites across the country as we collaborate with clients far and wide. We're best known for our work with the UK's major water and sewerage companies, but have expertise across several sectors, including:

- **Environmental engineering – with the Environment Agency and Coal Authority**
- **Highways – with local authorities**
- **Industrial building – with Rolls-Royce and Procter & Gamble**
- **Energy – with Magnox and Cadent Gas**

Our approach to business is one of collaboration: we've grown to the size we are because of repeat orders. We pride ourselves on being flexible and finding innovative solutions to our clients' challenges, applying the latest technology to drive efficiencies, reduce carbon and add social value.

"I first joined JN Bentley on a six-week civil engineering placement before returning as a graduate a couple of years later. The experience I gained on placement was really valuable: I got involved with completing site surveying and setting out and undertook the risk management and planning process. Since re-joining I've been given extra responsibility and am currently working on an important project spending my time completing civil engineering design, site engineering works and helping project lead two projects around the north west of London."

Sarah Biggins | JN Bentley graduate civil engineer

Our people

We can't do anything in this industry without people working together in great teams – which means that people are at the very heart of JN Bentley.

We offer opportunities across the civil engineering spectrum, with entry points including apprenticeships, placements and graduates, as well as experienced engineers.

Once on the books, our people then have the chance to gain valuable skills and experience right from day one. We use clear development pathways designed to help you maximise your potential – including routes to chartership. To help you get there you're given hands-on experience coupled with award winning development and training. And being part of the Mott MacDonald Group means the opportunities for development have never been greater.

Nothing gives us more satisfaction than seeing colleagues progress through JN Bentley which means we have a strong culture of promoting from within. Many of our senior leaders have progressed up through different civil engineering routes, showing positive 'can do' attitudes and drives to achieve.

Civil engineering is a dynamic, fast-moving industry and we'd love to see you become part of it with JN Bentley. To learn more:

www.jnbentley.co.uk

info@jnbentley.co.uk

01756 799425

Company	Website	Further information
HR Wallingford Ltd	hrwallingford.co.uk	hrwallingford.com/job-opportunities
Hydrock Consultants	hydrock.com	hydrock.com/careers
Hydrotec Consultants Ltd	hydrotech.uk.com	hydrotecltd.co.uk/contact
I & H Brown Limited	ihbrown.com	ihbrown.com/careers
IDOM Merebrook	merebrook.co.uk	merebrook.co.uk/careers/working-for-idom-merebrook
IKM Consulting Ltd	ikm.com	ikmconsulting.co.uk/ikm-careers
Independent Design House Ltd (IDH Design)	idh-design.co.uk	idh-design.co.uk/about/careers
Inertia Consulting Limited	inertiaconsulting.co.uk	inertiaconsulting.co.uk/contact-us
Intrafor Hong Kong Ltd	intrafor.com	intrafor.com/careers-school-relationships
Ironside Farrar Ltd	ironsidefarrar.com	ironsidefarrar.com/vacancy
ISG Construction Limited	isgltd.com	isgltd.com/en/careers/join-us
Isle of Man Government: Department of Infrastructure	gov.im	jobcentre@gov.im
J Murphy & Sons Ltd	murphygroup.co.uk	murphygroup.com/careers
J N Bentley Ltd	mottmacbentley.co.uk/introducing-jn-bentley	info@jnbentley.co.uk
J Reddington Ltd Group (JRL)	jrlgroup.co.uk	peoplebank.com/pb3/corporate/jrlgroup/page.php?p_page=careers

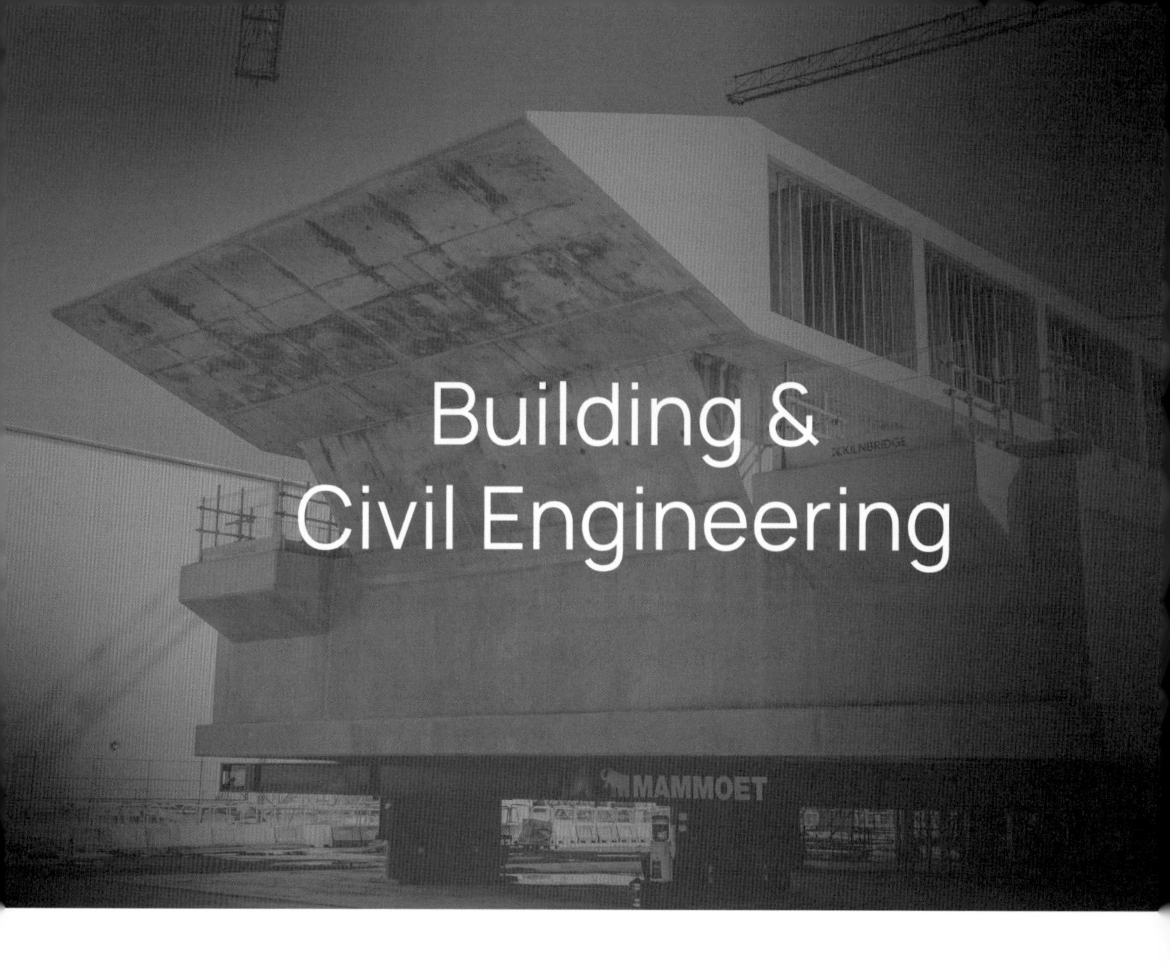
Building &
Civil Engineering
KILNBRIDGE
MAMMOET

Reinforced Concrete
Structures

KILNBRIDGE
HS2

Company	Website	Further information
J T Mackley & Co Limited	mackley.co.uk	mackley.co.uk/careers
Jackson Civil Engineering	jackson-civils.co.uk	jackson-civils.co.uk/careers
Jacobs UK Limited	**jacobs.com**	**careers.jacobs.com**
JBA Consulting (Jeremy Benn Associates)	jbaconsulting.com	jbaconsulting.com/work-for-us/careers
JDL Consultants Limited	jdlconsultants.co.uk	jdlconsultants.co.uk/vacancies
JNP Group Consulting Engineers Ltd	jnpgroup.co.uk	jnpgroup.co.uk/why-us/careers
John F Hunt Limited	johnfhunt.co.uk	johnfhunt.co.uk/contact-us
John Grimes Partnership Ltd	johngrimes.co.uk	johngrimes.co.uk/careers
John Sisk & Son Ltd	sisk.co.uk	johnsiskandson.com/uk/our-people/early-careers
Jones Bros Ruthin (Civil Engineering) Co Ltd	jones-bros.com	jonesbros-careers.com
JP Structural Design Ltd	jpstructural.co.uk	jpstructural.co.uk/contact
JPP Consulting	jppuk.net	jppuk.net/work-with-us
JRC Consulting Engineers Ltd	jrcconsulting.co.uk	jrcconsulting.co.uk/join-jrc-consulting
Jubb Consulting Engineers	jubb.uk.com	k.williams@jubb.uk.com
KBR	kbr.com	careers.kbr.com
Keller Ltd	keller.co.uk	keller.co.uk/careers
Kent PLC	kentplc.com	kentplc.com/careers
KEO International Consultants	keoic.com	www.keoic.com/graduate-internship-careers
Kier Group plc	**kier.co.uk**	**kier.co.uk/careers**
Kilnbridge Construction Services Ltd	kilnbridge.com	recruitment@kilnbridge.com
Kingscote Design Ltd	morrisroe.co.uk/companies/kingscote-design	morrisroe.co.uk/contact
Kirklees Council	kirklees.gov.uk	jobs@kirklees.gov.uk
Knights Brown	raymondbrowngroup.co.uk	knightsbrown.co.uk/graduates-trainees-and-apprentices
Lagan Specialist Contracting Group	laganconstruction.com	laganscg.com/careers
Laing O'Rourke plc	laingorourke.com	careers.laingorourke.com
Lancashire County Council	lancashire.gov.uk	lancashire.gov.uk/jobs
Leeds City Council	leeds.gov.uk	jobs.leeds.gov.uk
Leicestershire County Council	leicestershire.gov.uk	leicestershire.gov.uk/jobs-and-volunteering
Lend Lease Ltd	lendlease.com	lendlease.com/uk/careers
Lincolnshire County Council	lincolnshire.gov.uk	lincolnshire.gov.uk/jobs-careers
London Borough of Hammersmith & Fulham	lbhf.gov.uk	lbhf.gov.uk/jobs
London Borough of Waltham Forest	walthamforest.gov.uk	hradmin@walthamforest.gov.uk
LONG Engineering Limited	longengineering.co.uk	longengineering.co.uk/contact-us
Lorien Engineering Solutions	lorienengineering.com/en-gb	lorienengineering.com/en-gb/careers/working-at-lorien

Company	Website	Further information
M Group Services	mgroupservices.com	mgsworkwithus.com/company/M-Group-Services-Group-Support
M & D Foundations	mdfoundations.com	mdfoundations.com/careers
Mabey Bridge Ltd	mabeybridge.co.uk	mabeybridge.com/contact-us
Mabey Hire Ltd	mabey.co.uk	mabeyhire.co.uk/join-us
Mace Group Ltd	macegroup.com	careers.macegroup.com/early-careers
Magnox Limited	gov.uk/government/organisations/magnox-ltd	careers.magnoxsites.com
Manchester Airports Group	magairports.com	magairports.com/careers
Mann Williams	mannwilliams.co.uk	mannwilliams.co.uk/careers
Markides Associates Ltd	markidesassociates.co.uk	markidesassociates.co.uk/careers
Mason Clark	masonclark.co.uk	masonclark.co.uk/recruitment
McAdam Design	mcadamdesign.co.uk	mcadamdesign.co.uk/careers
McGee Group Ltd	mcgee.co.uk	mcgee.co.uk/join-us/why-mcgee
McLaughlin & Harvey Ltd	mclaughlinandharvey.com	mclh.co.uk/careers/our-opportunities
Meinhardt (Hong Kong) Limited	meinhardtgroup.com/offices/china	meinhardtgroup.com/careers
Meinhardt UK Ltd	meinhardt.co.uk	meinhardt.co.uk/careers
Metis Consultants Ltd	metisengineering.com	metisconsultants.co.uk/careers
MGF Design Services Ltd	mgf.co.uk	mgf.co.uk/contact-us
MHA Structural Design Ltd	mhastructuraldesign.com	mhastructuraldesign.com/contact
MHB Consultants	mhbconsultants.com	mhbconsultants.com/careers
Milk Architecture and Design Ltd	splashofmilk.com	splashofmilk.com/contact
Milner Associates	milnerassociates.co.uk	milnerassociates.co.uk/work-for-us
Ministry of Development, Brunei	mod.gov.bn	info@mod.gov.bn
Morgan Sindall Infrastructure	**morgansindall.com**	**morgansindallinfrastructure.com/join-our-team/working-with-us**
Morgan Structural Limited	morganstructural.co.uk	morganstructural.co.uk/we-are-recruiting
Mott MacDonald Limited	mottmac.com	mottmac.com/careers
Multiconsult	multiconsultgroup.com	multiconsultgroup.com/career
Multiplex Construction Europe Ltd	multiplex.global	multiplex.global/careers
MWH Treatment	mwhtreatment.com	careers.mwhtreatment.com/vacancies
Narro Associates	davidnarro.co.uk	narroassociates.com/careers
National Highways	nationalhighways.co.uk	careers.nationalhighways.co.uk
Natta Building Ltd	natta.co.uk	natta.co.uk/careers
Natural Resources Wales	naturalresourceswales.gov.uk	naturalresources.wales/about-us/jobs-and-placements/jobs
Network Rail	networkrail.co.uk	networkrail.co.uk/careers

Company	Website	Further information
Newcastle City Council	newcastle.gov.uk	newcastle.gov.uk/services/jobs-and-careers
Newtecnic Ltd	newtecnic.com	newtecnic.com/contact
Nexus	nexus.org.uk	recruitment@nexus.org
NIRAS Group (UK) Ltd	niras.com	niras.com/jobs
Norfolk County Council	norfolk.gov.uk	careers.norfolk.gov.uk
North Wales Joint Councils	gwynedd.gov.uk	gwynedd.llyw.cymru/en/Council/Jobs/Working-for-us
North Yorkshire County Council	northyorks.gov.uk	northyorks.gov.uk/jobs-and-careers
Northumberland County Council	northumberland.gov.uk	northeastjobs.org.uk
OCSC (Multidisciplinary Consulting Engineers)	ocsc.ie	ocsc.ie/contact-us/uk
Octavius Infrastructure Ltd.	octaviusinfrastructure.co.uk	octaviusinfrastructure.co.uk/careers
Odyssey Markides	odysseyconsult.co.uk	odysseyconsult.co.uk/careers
Offshore Wind Consultants Limited (OWC)	owcltd.com	abl-group.pinpointhq.com/owc
One Creative Environments	oneltd.com	oneltd.com/careers
ONLYgeotechnics Limited	onlygeotechnics.com	onlygeotechnics.com/contactus

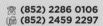

Company	Website	Further information
Opus International (M) Berhad - UEM Edgenta	uemedgenta.com/core-sectors/opus-consultants	uemedgenta.com/about-us/contact-us
P & S Consulting Engineers Ltd	pascoe-ltd.co.uk	pascoe-ltd.co.uk/contacts
Patrick Parsons Limited	patrickparsons.co.uk	patrickparsons.co.uk/join-our-team/
PCA Consulting Engineers	pcaconsulting.uk	pcaconsulting.uk/careers
PCE Limited	pceltd.co.uk	pceltd.co.uk/careers
PDMA Consulting Engineers Limited	pdmace.com	info@pdmace.com
Pell Frischmann Consultants Ltd	pellfrischmann.com	pellfrischmann.com/career-opportunities
Pembrokeshire County Council	pembrokeshire.gov.uk	pembrokeshire.gov.uk/jobs-and-careers
Perega	perega.co.uk	perega.co.uk/about-us/careers
PFA Consulting Ltd	pfaplc.com	pfaplc.com/careers
PHG Consulting Engineers Ltd	phg-consulting.co.uk	enquiries@phg-consulting.com
Phoenix Design Partnership	phoenixdp.co.uk	phoenixdp.co.uk/contact-phoenix
Pick Everard	pickeverard.co.uk	pickeverard.co.uk/careers
Pinnacle Consulting Engineers	pinnacleconsultingengineers.com	pinnacleconsultingengineers.com/careers
PJA Civil Engineering Ltd	pja.co.uk	pja.co.uk/about/join-us
Portsmouth City Council	portsmouth.gov.uk	careers.portsmouth.gov.uk
Price and Myers	pricemyers.com	pricemyers.com/contact/work-with-us
Project Centre Limited	projectcentre.co.uk	marstonholdings.co.uk/projectcentre/careers
QuadConsult Ltd	quadconsult.co.uk	quadconsult.co.uk/careers
R J McLeod (Contractors) Limited	rjmcleod.co.uk	rjmcleod.co.uk/careers
R&W Civil Engineering	rwcivilengineering.co.uk	rwcivilengineering.co.uk/careers
RAB Consultants	rabconsultants.co.uk	rabconsultants.co.uk/careers
Ramboll UK	ramboll.co.uk	uk.ramboll.com/careers
Ramsay & Chalmers	ramsaychalmers.co.uk	ramsaychalmers.co.uk/recruitment
Rappor Consultants Ltd	rappor.co.uk	rappor.co.uk/careers
Red7 Marine Limited	red7marine.co.uk	red7marine.co.uk/marine-recruitment
Rendel Limited	rendel-ltd.com	rendel-ltd.com/careers
Reno Pipe Construction (HK) Limited	renopipe.com.hk	renopipe.com.hk/people
RGA Consulting Engineers Ltd	rga.co.je	rga.co.je/recruitment-structural-engineer-jobs
Richard Jackson Limited	richardjackson.uk.com	richardjackson.uk.com/careers
Richter Associates Ltd	richter.global	careers.richter.global/vacancies
Ridge & Partners LLP	ridge.co.uk	ridge.co.uk/careers
Ringway Jacobs (JV Eurovia)	ringway-jacobs.co.uk	ringwayjacobscareers.co.uk
RMD Kwikform	rmdkwikform.com	rmdkwikform.com/careers

Company	Website	Further information
Roads & Transport Authority (RTA Dubai)	rta.ae	jobs.dubaicareers.ae/careersection/
Robert Bird and Partners Limited	robertbird.com	robertbird.com/careers
Robert Walpole and Partners	robertwalpole.co.uk	robertwalpole.co.uk/careers
Robert West Consulting	robertwest.co.uk	robertwest.co.uk/careers/vacancies
Robertson Construction Group Ltd	robertson.co.uk	robertson.co.uk/careers
Rodgers Leask Ltd	rogersleask.com	rodgersleask.com/careers
Roger Bullivant Limited	roger-bullivant.co.uk	roger-bullivant.co.uk/about/careers
Roughan & O'Donovan	rod.ie	rod.ie/careers/job-vacancies
Royal Haskoning DHV (RHDHV)	royalhaskoning.com	royalhaskoningdhv.com/en/careers
Royal School of Military Engineering (RSME)	army.mod.uk/who-we-are/corps-regiments-and-units/corps-of-royal-engineers/1-royal-school-of-military-engineering-regiment	
RPS	rpsgroup.com	rpsgroup.com/careers
RWO Associates	rwo.group	rwo.group/contact
Salford City Council	salford.gov.uk	salford.gov.uk/jobs-skills-and-work
Sanderson Watts Associates Ltd	sandersonwatts.com	sandersonwatts.com/careers
Scott White and Hookins LLP	swh.co.uk	swh.co.uk/careers
Scottish Water	scottishwater.co.uk	scottishwater.co.uk/about-us/careers
Sellafield Ltd	gov.uk/government/organisations/sellafield-ltd	careers.sellafieldsite.co.uk
Severfield UK Ltd	severfield.com	severfield.com/careers
Severn Trent Water Limited	severntrent.co.uk	stwater.co.uk/search/?q=careers
Shell	shell.com	shell.co.uk/careers
Simpson Associates Consulting Engineers LLP	simpsoneng.com	simpsoneng.com/career-opportunities
Sir Robert McAlpine Ltd (SRM)	**srm.com**	**srm.com/careers**
Skanska UK	**skanska.co.uk**	**skanska.co.uk/about-skanska/careers**
SLC Rail	slcrail.com	slcrail.com/join-our-team
SLR Consulting Ltd	slrconsulting.com	slrconsulting.com/en/careers
Smith and Wallwork	smithandwallwork.com	smithandwallwork.com/recruitment
Gateley Smithers Purslow Limited	smitherspurslow.com	smitherspurslow.com/careers
SNC LAVALIN	snclavalin.com/en	careers.snclavalin.com
Somerset County Council	somerset.gov.uk	somerset.gov.uk/jobs-and-careers
South East Water Ltd	southeastwater.co.uk	southeastwater.co.uk/about/careers
South Gloucestershire Council	southglos.gov.uk	southglos.gov.uk/jobs-and-careers
South Hams District Council & West Devon Borough Council	southhams.gov.uk	jobs@swdevon.gov.uk
Southern Harbour Limited	southernharbour.net	southernharbour.net/contact

Company	Website	Further information
Southern Water (Nr) Holdings Ltd	southernwater.co.uk	southernwater.co.uk/careers
SSE	sse.com	sse.co.uk/about-us/careers
Staffordshire County Council	staffordshire.gov.uk	staffordshire.gov.uk/Jobs-and-careers/Jobs
Stantec	stantec.com	stantec.com/uk/careers
States of Guernsey	gov.gg	gov.gg/careers
States of Jersey	gov.je	recruitment@gov.je
Steel Construction Institute	steel-sci.org	reception@steel-sci.com
Stirling Maynard	stirlingmaynard.com	stirlingmaynard.com/careers
Stockport Metropolitan Borough Council	stockport.gov.uk	stockport.gov.uk/topic/jobs-training-and-skills
Stomor Ltd	stomor.com	stomor.com/careers
Story Contracting Ltd	storycontracting.com	storycontracting.com/jobs/why-work-for-us
Strabag UK Ltd	strabag.com	jobboerse.strabag.at/jobs-overview
Structa LLP	structa.co.uk	structa.co.uk/careers
Stuart Michael Associates Limited	stuartmichael.co.uk	stuartmichael.co.uk/careers-2
Studio One Consulting Limited	studio.engineering	studio.engineering/contact
Subsea 7	subsea7.com	subsea7.com/en/our-people
Such Salinger Peters Ltd	sspconsulting.co.uk	enquiries@sspconsulting.co.uk
Surrey County Council	surreycc.gov.uk	surreycc.gov.uk/jobs
Sustrans Limited	sustrans.org.uk	sustrans.org.uk/careers
Sutcliffe Projects	sutcliffe.co.uk	sutcliffe.co.uk/careers
Suttle Projects Ltd	suttles.co.uk	info@suttles.co.uk
Swann Engineering Group	swanngroupltd.com	swanngroupltd.com/category/careers
Swansea Council	swansea.gov.uk	swansea.gov.uk/councilcareers
Swanton Consulting Ltd	swantonconsulting.co.uk	swantonconsulting.co.uk/contact-us/careers
Sweco UK Limited	sweco.co.uk	sweco.co.uk/careers
Symmons Madge Associates Ltd	symmonsmadge.co.uk	admin@symmonsmadge.co.uk
Systra UK	systra.co.uk	systra.co.uk/en/systra/contact
T&G Limited	tag.je	tag.je/careers/t-g-jersey
Taranto Ltd	taranto.co.uk	taranto.co.uk/contact
Tata Steel Strip Products UK	tatasteeleurope.com	recruitmentuk@tatasteeleurope.com
Taylor & Boyd LLP	taylor-boyd.co.uk	taylor-boyd.com/careers
Teignbridge District Council	teignbridge.gov.uk	teignbridge.gov.uk/jobs-and-careers
Tetra Tech	tetratech.com	tetratech.com/en/careers
Thames Water Utilities Ltd	thames-water.com	thameswater.co.uk/about-us/careers
The Fletcher Construction Company Ltd	fletcherconstruction.co.nz	fletcherconstruction.co.nz/people-and-careers

Company	Website	Further information
The Highland Council	highland.gov.uk	highland.gov.uk/info/1084/jobs_and_careers/333/working_for_us
The Humber Bridge Board	humberbridge.co.uk	humberbridge.co.uk/humberbridge/recruitment
The Morton Partnership Ltd	themortonpartnership.co.uk	themortonpartnership.co.uk/about/the-team
The Spencer Group	thespencergroup.co.uk	thespencergroup.co.uk/vacancies
UtilityINFO Limited (TUG Group)	tug.hk	utilityinfo.com.hk/career
Thomas Consulting Limited	thomasconsulting.co.uk	thomasconsulting.co.uk/careers
Thornton Tomasetti	thorntontomasetti.com	thorntontomasetti.com/careers
Tilbury Douglas Construction Limited	tilburydouglas.co.uk	tilburydouglas.co.uk/careers
TOBIN Consulting Engineers	tobinconsultingengineers.com	tobinconsultingengineers.com/careers
Tony Gee & Partners LLP	**tgp.co.uk**	**tonygee.com/careers**
TopBond PLC	topbond.co.uk	topbond.co.uk/careers
Translink	translink.co.uk	translink.co.uk/workwithus
Transport for London (TfL)	tfl.gov.uk	tfl.gov.uk/corporate/careers
Transport for Wales	tfw.wales	tfw.wales/info-for/job-hunters
Transport Scotland	transport.gov.scot	transport.gov.scot/careers
Trant Engineering Ltd	trant.co.uk	trant.co.uk/careers/vacancies
Tully De'Ath Consultants Limited	tullydeath.com	tullydeath.com/contact-us/join-us
Turner & Townsend	turnerandtownsend.com	turnerandtownsend.com/en/careers
United Utilities PLC	unitedutilities.com	unitedutilities.com/corporate/careers
University of Bristol	bristol.ac.uk	recruitment@bristol.ac.uk
United Nations Office for Project Services (UNOPS)	unops.org	jobs.unops.org/pages/viewvacancy/valisting
Van Elle Ltd	van-elle.co.uk	van-elle.co.uk/careers
VEDA Associates Ltd.	veda.co.uk	veda.co.uk/careers
VGC Group	vgcgroup.co.uk	vgcgroup.co.uk/jobs
VINCI Construction Grands Projets Hong Kong	vinci-construction-projets.com	vinci-construction-projets.com/en/contact
VINCI Construction UK Ltd	vinciconstruction.co.uk	vinciconstruction.co.uk/work-with-us/achieve-your-potential
VolkerFitzpatrick	volkerfitzpatrick.co.uk	enquiries@volkerfitzpatrick.co.uk
VolkerRail	volkerrail.co.uk	volkerrail.co.uk/en/careers
VolkerStevin Ltd	volkerstevin.co.uk	volkerstevin.co.uk/en/careers
Vp Plc Groundforce Shorco	vpgroundforce.com	vpgroundforce.com/gb/footer-links/useful-links/contact-us
Waldeck Associates Ltd	waldeckconsulting.com	waldeckconsulting.com/careers

Company	Website	Further information
VolkerFitzpatrick volkerfitzpatrick.co.uk enquiries@volkerfitzpatrick.co.uk		As one of the UK's leading engineering and construction companies, VolkerFitzpatrick provides specialist building, civil engineering and rail expertise to a range of markets to provide a truly integrated service to clients. We have proven experience and deliver successful projects in a wide range of industries, including highways commercial, industrial, education, rail infrastructure and depots, airports and energy. Founded in 1921, the company has grown to become one of the top contractors in the UK. VolkerFitzpatrick's success is based on ensuring it has a sound understanding of the client's vision, which is then delivered on time and to budget. VolkerFitzpatrick is part of VolkerWessels UK, a multi-disciplinary construction and civil engineering group with a turnover of £1.15 billion. VolkerWessels UK employs more than 3,500 staff in five operating companies. Contact us for more information regarding our award winning Degree apprentice and graduate programme at 01992 305 000 or email **Sally.Hill@volkerfitzpatrick.co.uk.**
Waldon Telecom Ltd	waldontelecom.com	mgsworkwithus.com/company/waldon-telecom/
Walsh Associates	walshandassociates.co.uk	enquiries@walshandassociates.co.uk
Walters UK Ltd	walters-uk.co.uk	walters-group.co.uk/careers
Ward and Burke Construction	wardandburke.com	wardandburke.com/careers
WARD Associates (Consulting Engineers) Ltd.	wardac.co.uk	wardac.co.uk/#section-contact
Wardell Armstrong LLP	wardell-armstrong.com	wardell-armstrong.com/en/careers
Warwickshire County Council	warwickshire.gov.uk	warwickshire.gov.uk/jobs
Waterco Ltd	waterco.co.uk	waterco.co.uk/contact
Waterman Aspen Limited	watermanaspen.co.uk	watermanaspen.co.uk/careers
Waterman Group	watermangroup.com	watermangroup.com/careers/vacancies
Wates Construction Ltd	wates.co.uk	wates.co.uk/careers
WDR & RT Taggart	taggarts.uk	info@taggarts.uk
Webb Yates Engineers	webbyates.com	webbyates.com/careers
Wentworth House Partnership (Keltbray)	wentworth-house.co.uk	wentworth-house.co.uk/careers
Wessex Water	wessexwater.co.uk	wessexwater.co.uk/careers
West Berkshire Council	westberks.gov.uk	jobs.westberks.gov.uk
Westlakes Engineering Ltd	westlakes-engineering.co.uk	westlakes.co.uk/contact-us
Whitby Wood	whitbywood.com	whitbywood.com/join-us
Whitfield Construction Services		whitfieldconstructionservices.co.uk/careers
WHP Telecoms Ltd	whptelecoms.com	whptelecoms.com/careers
Wigan Council	wigan.gov.uk	wigan.gov.uk/Resident/Jobs-Careers
Wilde Consultants Ltd	wilde.uk.com	wildecivil.co.uk/vacancies

Company	Website	Further information
Wills Bros Civil Engineering Ltd	willsbros.com	willsbros.com/careers
Wiltshire Council	wiltshire.gov.uk	jobs.wiltshire.gov.uk
Wing Hing Construction Co. Ltd	wh-cwf.com/en/WH	info@whconst.com.hk
Wings & Associates Consulting Engineers Ltd	wingsconsulting.com.hk	info@whconst.com.hk
Winvic Construction Ltd	winvic.co.uk	winvic.co.uk/work-at-winvic
Wokingham Borough Council	wokingham.gov.uk	wokingham.gov.uk/jobs-and-careers
Wood PLC	woodplc.com	careers.woodplc.com
Wood Thilsted Partners	woodthilsted.com	woodthilsted.com/careers
Woolgar Hunter Limited	woolgarhunter.com	woolgarhunter.com/careers
Worley Europe Limited	worleyparsons.com	worley.com/careers
Wormald Burrows Partnership Ltd	wormburp.com	engineer@wormburp.com
WSP	**wsp.com**	**wsp.com/en-gb/careers**
WSP RE&I LTD	wsp.com	wsp.com/en-gb/careers
Xeiad	xeiad.com	xeiad.com/what-we-do/careers
Yorkshire Water	yorkshirewater.com	yorkshirewater.com/careers/vacancies
YWL Engineering Limited	ywlgroup.com	ywlgroup.com/ywl-wp/career
Xeiad	xeiad.com	xeiad.com/what-we-do/careers
Yorkshire Water	yorkshirewater.com	yorkshirewater.com/careers/vacancies
YWL Engineering Limited	ywlgroup.com	ywlgroup.com/ywl-wp/career

ICE corporate partners are listed in bold

This list is correct as of 1st December 2022

Image Credits

Page 7:	Train station platform. William Barton/Shutterstock.
Page 8:	Aerial landscape of the harbor in Wladyslawowo by the Baltic Sea at summer, Poland. Patryk Kosmider/Shutterstock.
Page 12:	Crosby beach near Liverpool. Jason Wells/Shutterstock.
Page 13:	Eastern Scheldt storm surge barrier, Netherlands. Jacco Vasseur/Shutterstock.
Page 14:	Shutter.B/Shutterstock.
Page 16:	Vasco da Gama Bridge, Portugal. Piotr Szczepanek/Shutterstock.
Page 17:	One Great George Street. Institution of Civil Engineers.
Page 20:	Romeo Pj/Shutterstock.
Page 21:	Fizkes/Shutterstock.
Page 24:	Traffic crosses a busy intersection in Shibuya, Tokyo, Japan. TierneyMJ/Shutterstock.
Page 27:	The Shard, London. Andy Shiels/Shutterstock.
Page 29:	iJeab/Shutterstock.
Page 30:	Millau Viaduct, France. Lenush/Shutterstock.
Page 33:	Old Trafford football stadium, Manchester, UK. Alex Motoc/Unsplash.
Page 36:	Artens/Shutterstock.
Page 40:	40 Leadenhall Street, London. Mace.
Page 44:	Circular Quay and Opera House, Sydney, Australia. Ira Sokolovskaya/Shutterstock.
Page 45:	Golden Gate Bridge, San Francisco CA USA. titipongpwl/Shutterstock.
Page 46:	1. The Menai suspension bridge designed by Thomas Telford and completed in 1826 to connect the mainland with the island of Anglesey Wales. David Pimborough/Shutterstock.
	2. San Francisco bay. Golden Gate Bridge with San Francisco city and bay bridge as background. Min C. Chiu/Shutterstock.
	3. Çanakkale Bridge. Zafer/ https://creativecommons.org/licenses/by-sa/4.0/deed.en.
Page 47:	Øresund Bridge between Denmark and Sweden. Ingus Kruklitis/Shutterstock.
Pages 49 and 50:	Gull Wing Bridge, Lowestoft, UK. Farrans Construction.
Page 51:	Metropol Parasol located in the old quarter of Seville, Spain. Irina Wilhauk/Shutterstock.
Page 52:	Canary Wharf Station, London, United Kingdom. Alexander London/Unsplash.
Page 55:	Soil sampling in the production of engineering-geological surveys. Artur_Sarkisyan/Shutterstock.
Page 56:	Odometer cells for soil testing in geotechnics laboratory. Skinfaxi/Shutterstock.
Page 57:	Construction site for big building foundation. Ioan Panaite/Shutterstock.
Page 58:	Landslide rocks blocking Santa Susana Pass Road in Los Angeles, California. trekandshoot/Shutterstock.
Page 62:	xdoganbor/Shutterstock.
Page 63:	SvedOliver/Shutterstock.
Page 66:	Have a nice day Photo/Shutterstock.
Page 67:	sabthai/Shutterstock.
Page 69:	Above the Channel Tunnel Entrance by Chris Heaton. This file is licensed under the Creative Commons Attribution-Share Alike 4.0 International license.

Page 69:	This photo was taken by Florian Fèvre from Mobilys. This file is licensed under the Creative Commons Attribution-Share Alike 4.0 International license.
Page 70:	BigPixel Photo/Shutterstock.
Page 73:	Thames Barrier, London, UK.
Page 74:	Hydroelectric Dam. Parilov/Shutterstock.
Page 75:	Hoover Dam. tvamvakinos/Shutterstock.
Page 76:	Amorn Suriyan/Shutterstock.
Page 77:	Thames Tideway Tunnel Boring Machine Rachel named after Rachel Parsons, first President and founder of the Women's Engineering Society. MyrtleGal/ https://creativecommons.org/licenses/by-sa/4.0/deed.en.
Page 78:	chekart/Shutterstock.
Page 79:	Marco de Benedictis/Shutterstock.
Page 82:	M-Production/Shutterstock.
Pages 83 to 86:	Loch Ness: water treatment works, Scotland. Galliford Try.
Page 91:	Coastal defences on Marine Drive Scarborough, England, UK. Atlaspix/Shutterstock.
Page 92:	Delma Port, Abu Dhabi, UAE.
Page 93:	Port of Liverpool Seaforth Docks on the River Mersey, Merseyside, UK. Neil Mitchell/Shutterstock.
Page 94:	Rossall Coastal Defence Scheme, Fylde coast, UK. Balfour Beatty.
Page 96:	Great Yarmouth Tidal Defences, JN Bentley.
Page 98:	North Sea Terminal Bremerhaven, Germany.
Page 100:	Aerial view on Palm Jumeira island in Dubai, UAE. Funny Solution Studio/Shutterstock.
Page 103:	Concrete interlocking tetrapods. philip openshaw/Shutterstock.
Page 104:	Keppel Port, Singapore.
Page 105:	Sea defences overlooking Morecambe Bay, UK. Neil Mitchell/Shutterstock.
Page 107:	High Speed Train. zhang sheng/Shutterstock.
Page 109:	Kluuvi District, Helsinki, Finland. Grisha Bruev/Shutterstock.
Page 110:	AUUSanAKUL/Shutterstock.
Page 111:	Airport. Jaromir Chalabala/Shutterstock.
Page 112:	Modern buildings with empty road, Tianjin, China. ssguy/Shutterstock.
Page 115:	Forth Bridge Queensferry Crossing under construction. Marten_House/Shutterstock.
Page 116:	Gorodenkoff/Shutterstock.
Page 117:	Elizabeth Line. Kevin Grieve/Unsplash.
Page 120:	Airport Terminal. creativemarc/Shutterstock.
Page 123:	UK cycle lane, London. Sampajano_Anizza/Shutterstock.
Page 124:	Ekaterina Pokrovsky/Shutterstock.
Page 127:	London Stadium in Queen Elizabeth park, London, United Kingdom. Aerial-motion/Shutterstock.
Page 129:	Park Guell in Barcelona, Spain. Georgios Tsichlis/Shutterstock.
Page 130:	Tianjin china. ssguy/Shutterstock.
Page 131:	Light rail Metrolink tram in the city center of Manchester, UK. Madrugada Verde/Shutterstock.
Page 133:	Renewable energy. metamorworks/Shutterstock.
Page 135:	1. Hydroelectric dam. Evgeny_V/Shutterstock.
	2. Rampion Offshore Wind Farm, UK.

Page 137:	Oleksii Sidorov/Shutterstock.
Page 138:	Ghazi-Barotha hydropower project - Ghazi-Barotha, Pakistan.
Page 144 to 146:	Sellafield. Sellafield Ltd.
Page 151:	City park and skyline.
Page 152:	Queen Elizabeth Olympic Park/Amanda Slater. This file is licensed under the Creative Commons Attribution-Share Alike 2.0 Generic license.
Page 154:	Little Island park at Pier 55 in New York. Creative Family/Shutterstock.
Page 155:	1. The Chicago River and downtown Chicago skyline USA. f11photo/Shutterstock.
	2. Urban Planning. Rawpixel.com/Shutterstock.
Page 156:	Songdo Central Park in Songdo District, Incheon South Korea. PKphotograph/Shutterstock.
Page 162:	Nahulingo, El Salvador. Milosz Reterski/ Bridges to Prosperity. This file is licensed under the Creative Commons Attribution-Share Alike 3.0 Unported license.
Page 164:	Minigo Bridge, Rwanda. Balfour Beatty.
Page 166:	Road in Annapurna range of Himalayan mountains. Sandra Foyt/Shutterstock.

Acknowledgements

Alan Macleay, EPCI Solutions & Integrated Projects
Alastair Smyth, Byrne Bros
Alex Radcliffe, Beca
Alexander Hay, Southern Harbour Ltd
Bill Hewlett, British Board of Agrément
Caleb Quaye, Arup
Carlos Rueda, Arcadis
Charlotte Flower, Mott MacDonald
Chris McCall, Retired
Dr Natalie Wride, Rendel Limited
Eva Johnson, Arup
Firoozeh Moghaddam, DNV
Georgina Seely, Thames Water
Helen Bradford, J N Bentley
Hilary Shields, Arup
Jack Graham, Waterman Aspen
Jane Wright, Transport for London
Jo Parker, Institution of Civil Engineers
Julian Phatarfod, Arup
June Chen, Wing Hing Construction Company Ltd
Kathryn Cavanagh, Sellafield
Kelly McNee, Amee
Kevin Drain, Atkins
Kiera Young, Mace Group
Lucy Cryer, Arup
Majid Jamil, Kier Group
Marisa Ackurst, Zutari (Pty) Ltd
Olufemi Ajayi, Arcadis
Pankaj Garj, Atkins
Paula McMahon, Sir Robert McAlpine
Philip Pascal, Member of Energy Committee of WFEO
Pierpaolo Avanzi, Mace
Richard Scantlebury, Hewson Consulting
Rob Dunn, J N Bentley
Scott Magee, Arup
Stephen Bennett, Arup
Suzi Campbell, Farrans
Usman Ali, Arcadis
Zoe Sturgess, Jacobs